P9-BBQ-416

'Religion without science is blind. Science without religion is lame.'

Albert Einstein

By the same author

The Runaway Universe
Other Worlds
The Edge of Infinity: Naked Singularities and the Destruction of Spacetime

Student texts
Space and time in the modern universe
The forces of nature
The search for gravity waves
The accidental universe

Technical
The physics of time asymmetry
Quantum fields in curved space (with N.D. Birrell)

Paul Davies

GOD *and the* New Physics

A TOUCHSTONE BOOK
Published by Simon & Schuster
New York London Toronto Sydney

First Touchstone Edition, 1984
Published by Simon & Schuster, Inc.

Rockefeller Center
1230 Avenue of the Americas
New York, New York 10020

Manufactured in the United States of America
Originally published in England in 1983
by J. M. Dent & Sons Ltd.

10 9 8 7 6 5 4 3 2 1
30 29 28 27 Pbk.

Library of Congress Cataloging in Publication Data

Davies, Paul, date.
God and the new physics.

1. Physics—Religious aspects—Christianity.
2. Religion and science—1946- . I. Title.
BL265.P4D38 1983 261.5′5 83-14866
ISBN 0-671-47688-2
ISBN 0-671-52806-8 Pbk.

Contents

Acknowledgements

I should like to give special thanks to Dr. John Barrow of the University of Sussex, whose detailed comments have much improved the presentation of this book. The subject matter has also provoked some lively coffee-time discussions in my own department, and I have found conversation with Dr. Stephen Bedding, Mr. Kerry Hinton, Dr. J. Pfautsch, Dr. Stephen Unwin and Mr. William Walker very fruitful.

The author and publishers wish to thank: Faber and Faber Ltd for permission to quote from 'The Expanding Universe' by Norman Nicholson in *The Pot Geranium*; Harvester Press Ltd for permission to quote from *Gödel, Escher, Bach* by D.R. Hofstadter and *The Mind's I* by D.R. Hofstadter and D.C. Dennett; Methuen London Ltd for permission to quote from *Summa Theologiae*, Vol I: *Christian Theology* by St Thomas Aquinas, edited by Thomas Gilby; Richard P. Feynman to quote from his book *The Character of Physical Law*; Pergamon Press Ltd for permission to quote from Sir Herman Bondi, 'Religion is a good thing' in *Living Truths*, edited by Ronald Duncan and Miranda Weston-Smith.

Preface

Over fifty years ago something strange happened in physical science. Bizarre and stunning new ideas about space and time, mind and matter, erupted among the scientific community. Only now are these ideas beginning to reach the general public. Concepts that have intrigued and inspired physicists themselves for two generations are at last gaining the attention of ordinary people, who never suspected that a major revolution in human thought had occurred. The new physics has come of age.

In the first quarter of this century two momentous theories were proposed: the theory of relativity and the quantum theory. From them sprang most of twentieth-century physics. But the new physics soon revealed more than simply a better model of the physical world. Physicists began to realize that their discoveries demanded a radical reformulation of the most fundamental aspects of reality. They learned to approach their subject in totally unexpected and novel ways that seemed to turn commonsense on its head and find closer accord with mysticism than materialism.

The fruits of this revolution are only now starting to be plucked by philosophers and theologians. Many ordinary people too, searching for a deeper meaning behind their lives, find their beliefs about the world very much in tune with the new physics. The physicist's outlook is even finding sympathy with psychologists and sociologists, especially those who advocate a holistic approach to their subjects.

In giving lectures and talks on modern physics I have discerned a growing feeling that fundamental physics is pointing the way to a new appreciation of man and his place in the universe. Deep questions of existence — How did the universe begin and how will it end? What is matter? What is life? What is mind? — are not new. What is new is that we may at last be on the verge of answering them. This astonishing prospect stems from some spectacular recent advances in physical science — not only the new physics, but its close relative, the new cosmology.

For the first time, a unified description of all creation could be within our grasp. No scientific problem is more fundamental or more daunting than the puzzle of how the universe came into being. Could this have happened without any supernatural input? Quantum physics seems to provide a loophole to the age-old assumption that 'you can't get something for nothing'. Physicists are now talking about 'the self-creating universe': a cosmos that erupts into existence spontaneously, much as a subnuclear particle sometimes pops out of nowhere in certain high energy processes. The question of whether the details of this theory are right or wrong is not so very important. What matters is that it is now possible to conceive of a scientific explanation for all of creation. Has modern physics abolished God altogether?

This is not a book about religion. Rather, it is about the impact of the new physics on what were formerly religious issues. In particular, I make no attempt to discuss religious experiences or questions of morality. Nor is it a science book. It is a book *about* science and its wider implications. Inevitably, it is necessary here and there to explain some technicalities in careful detail, but I do not claim that the scientific discussions are either systematic or complete. The reader should not be deterred by the thought that he or she is in for some punishing mathematics or strings of specialist terminology. I have tried to avoid technical jargon as much as possible.

The book is primarily intended for the general reader, both aetheist and believer, with no previous knowledge of science. However, I hope that it also contains some material of real scholarly value. In particular, I do not believe that some of the very recent work on cosmology has previously come to the attention of philosophers and theologians.

The central theme of the book concerns what I call the Big Four questions of existence:

Why are the laws of nature what they are?

Why does the universe consist of the things it does?

How did those things arise?

How did the universe achieve its organization?

Towards the end of the book, tentative answers to these questions begin to emerge — answers based on the physicist's conception of nature. The answers may be totally wrong, but I believe that physics is uniquely placed to provide them. It may seem bizarre, but in my opinion science offers a surer path to God than religion. Right or wrong, the fact that science has actually advanced to the point where what were formerly religious questions can be seriously tackled, itself indicates the far-reaching consequences of the new physics.

Although I have endeavoured to exclude my own religious opinions throughout, my presentation of physics is inevitably a personal one. No doubt many of my colleagues would strongly disagree with the conclusions I attempt to draw. I respect their opinions. This is simply one man's perception of the universe; there are many others. My motivation for writing the book is that I am convinced there is more to the world than meets the eye.

Note on technical terms

The word 'billion' is used to mean one thousand million. Occasionally it is convenient to use the shorthand 'powers of ten' notation for very large or small numbers. For example, 10^6 denotes one million, 10^9 one billion, 10^{-6} one millionth, 10^{-9} one billionth.

1. Science and religion in a changing world

'The wise man regulates his conduct by the theories both of religion and science.'

J.B.S. Haldane

'But because I have been enjoined, by this Holy Office, altogether to abandon the false opinion which maintains that the Sun is the centre and immovable, and forbidden to hold, defend, or teach, the said false doctrine in any manner . . . I abjure, curse, and detest the said errors and heresies, and generally every other error and sect contrary to the said Holy Church . . .'

Galileo Galilei

Science and religion represent two great systems of human thought. For the majority of people on our planet, religion is the predominant influence over the conduct of their affairs. When science impinges on their lives, it does so not at the intellectual level, but practically, through technology.

In spite of the power of religious thought in the daily lives of the general public, most of our institutions are organized pragmatically, with religion, insomuch as it is included at all, relegated to a stylized role. Such is the constitutional position of the Church of England for example. There are exceptions: Ireland and Israel remain religious states in the legal sense, while the revival of militant Islam is, if anything, increasing the influence of religion in political and social decision making.

In the industrialized world, where the impact and success of science is most conspicuous, there has been a sharp decline in affiliation to the major traditional religious institutions. In Britain, only a tiny percentage of the population now attend church regularly. It would be a mistake, however, to conclude that declining church attendance can be directly attributed to the raised profile of science and technology. In their personal lives many people still hold deep beliefs about the world that could be classed as religious, even though they may have rejected, or at least ignored, the traditional Christian doctrines. And any scientist will verify that, if religion has been displaced from people's consciousness, it has certainly not been replaced by rational scientific thought. For science, despite its great impact on all our lives at the practical level, is as elusive and inaccessible to the general public as any exclusive religion.

More relevant to the decline of religion is the fact that science, through technology, has altered our lives so radically that the traditional religions may appear to lack the immediacy necessary to provide any real assistance in coping with contemporary personal and social problems. If the Church is largely ignored today it is not because science has finally won its age-old battle with religion, but because it has so radically reoriented our society that the biblical perspective of the world now seems largely irrelevant. As one television cynic recently remarked, few of our neighbours possess an ox or an ass for us to covet.

The world's major religions, founded on received wisdom and dogma, are rooted in the past and do not cope easily with changing times. Hastily discovered flexibility has enabled Christianity to incorporate some new features of modern thought, to the extent that today's Church leaders might well have appeared heretical to a Victorian; yet any comprehensive philosophy based on ancient concepts faces a hard task in adapting to the space age. As a result, many disillusioned believers have turned to 'fringe' religions that seem more in tune with the era of *Star Wars* and microchips. The huge rise in popularity of cults associated with UFOs, ESP, spirit contacts, scientology, transcendental meditation and other technology-based beliefs testifies to the continued persuasiveness of faith and dogma in a superficially rational and scientific society. For although these eccentric ideas have a scientific veneer, they are unashamedly irrational — 'cults of unreason', to use Christopher Evans's phrase from his book of the same title (Panther 1974). People turn to them not for intellectual enlightenment but for spiritual comfort in a harsh and uncertain world.

Science, then, has invaded our lives, our language and our religions, but not at the intellectual level. The vast majority of people do not understand scientific principles, nor are they interested. Science remains a sort of witchcraft, its practitioners regarded with a mixture of awe and suspicion. Browse through any bookshop. Books on science are usually catalogued under 'The Occult', and modern astronomy textbooks jostle *The Bermuda Triangle* and *Chariots of the Gods* for space on the shelves. Lip service may be paid to the importance of science and rational thought for ordering our society, but at the personal level most people still find religious doctrine more persuasive than scientific arguments.

We live in a world that, in spite of appearances, is still fundamentally religious. Ranging from countries like Iran and Saudi Arabia, where Islam remains the dominant social force, to the industrialized West, where religion has fragmented and diversified, occasionally into vague pseudo-scientific superstition, the search for a deeper meaning to life continues. Nor should that search be derided. Scientists also are searching for a meaning: by finding out more about the way the universe is put together and how it works, about the nature of life and consciousness, they can supply the raw material from which religious beliefs may be fashioned. To argue whether the date of the Creation was 4004 B.C. or 10,000 B.C. is irrelevant if scientific measurements reveal a 4½ billion-year-old Earth. No religion that bases its beliefs on demonstrably incorrect assumptions can expect to survive very long.

In this book we shall be looking at some of the very latest discoveries in fundamental science, and exploring their implications for religion. In many cases the old religious ideas are not so much disproved as transcended by modern science. By looking at the world from a different angle, scientists can provide fresh insights and new perspectives of Man and his place in the universe.

Both science and religion have two faces: the intellectual and the social. In both cases the social effects leave a lot to be desired. Science may have alleviated the miseries of disease and drudgery and provided an array of gadgetry for our entertainment and convenience, but it has also spawned horrific weapons of mass destruction and seriously degraded the quality of life. The impact of science on industrial society has been a mixed blessing.

On the other hand, organized religion comes off, if anything, even worse. Nobody denies the many individual cases of selfless devotion by religious community workers all over the world, but religion long ago became institutionalized, often concerning itself more with power and politics than with good and evil. Religious zeal has all too frequently been channelled into violent conflict, perverting man's normal tolerance and unleashing barbaric cruelty. Christian genocide of the South American native populations in the Middle Ages is one of the more dreadful examples, but the history of Europe generally is littered with the corpses of those slain because of minor doctrinal differences. Even in this so-called enlightened age, religious hatred and conflict fester all over the world. It is ironical that although most religions extol the virtues of love, peace and humility, it is all too often hatred, war and arrogance that characterize the history of the world's great religious organizations.

Many scientists are critical of organized religions, not because of their personal spiritual content, but for their perverting influence on otherwise decent human behaviour, especially when they involve themselves in power politics. The physicist Hermann Bondi is a harsh critic of religion, which he regards as a 'serious and habit-forming evil'. He cites as an example the excesses of the European witch-craze:

> In much of Christian Europe the godfearing used to burn old women suspected of being witches, an arduous duty they felt had been clearly put upon them by the Bible. The facts on witch burning are clear enough: First, faith made otherwise decent people commit acts of unspeakable horror, showing how ordinary and everyday feelings of human kindness and revulsion at cruelty can be and have been overruled by religious belief. Secondly, it exposes as utterly hollow the claim that religion sets an absolute and unchanging foundation for morality.[1]

Bondi claims that the ruthless power wielded by the Church and other religious institutions over the centuries leaves these organizations morally bankrupt.

Few would deny that religion remains, for all its pretentions, one of the most divisive forces in society. Whatever the good intentions of the faithful, the bloodstained history of religious conflict provides little evidence for universal standards of human morality among the major organized religions. Nor is there any reason to believe that love and consideration are lacking in those who do not belong to such organizations, or are even committed aetheists.

Of course, not all religious people are fanatical zealots. The vast majority of Christians today share a revulsion of religious conflict and deplore the Church's past involvement with torture, murder and suppression. But the outbreaks of spectacular violence and brutality in the name of God which still plague society today are not the only manifestations of the antisocial face of religion. Segregation in education and even habitation continues in supposedly civilized countries like Northern Ireland and Cyprus. Even within their own ranks, religious organizations often sanction prejudice, whether against women, racial minorities, homosexuals or whoever their leaders decree to be inferior. The status of women in Catholicism and Islam, or blacks in the South African Church, I find particularly offensive. Although many people would be appalled that their own religion might be described as vicious or intolerant, they will readily agree that the world's *other* religions have a lot to answer for.

This sad history of bigotry seems inevitably to result once religious organizations become institutionalized and constitutionalized, and has prompted a huge disaffection with established religion in the Western world. Many are turning instead to the so-called 'fringe' religions, in an attempt to find a less strident and more gentle route to spiritual fulfilment. There are, of course, a wide variety of new movements, some of which are still more intolerant and sinister than the traditional religions. But many emphasize the importance of mysticism and quiet inner exploration, as opposed to evangelical fervour, and so attract those people who are critical of the social and political impact of the established religions.

So much for the social side of religion. What of its intellectual content?

For the greater part of human history, men and women have turned to religion not only for moral guidance, but also for answers to the fundamental questions of existence. How was the universe created and how will it end? What is the origin of life and mankind? Only in the last few centuries has science begun to make its own contributions to such issues. The resulting clashes are well documented. From its origin with Galileo, Copernicus and Newton, through Darwin and Einstein, to the age of computers and high technology, modern science has cast a cold and sometimes threatening light on many deep-rooted religious beliefs. Accordingly, there has grown the feeling that science and religion are inherently incompatible and antagonistic. It is a belief encouraged by history. The early attempts by the Church to hold back the flood-gates of scientific advance have left a deep suspicion of religion among the scientific community. For their part, scientists have demolished a lot of cherished religious beliefs and have come to be regarded by many as faith-wreckers.

There is no doubt, however, about the success of the scientific method. Physics, the queen of sciences, has opened up vistas of human understanding that were unsuspected a few centuries ago. From the inner workings of the atom to the weird surrealism of the black hole, physics has enabled us to comprehend some of nature's darkest secrets and to gain control over many physical systems in our environment. The tremendous power of scientific reasoning is demonstrated daily in the many marvels of modern technology. It seems reasonable then, to have some confidence in the scientist's world-view also.

The scientist and the theologian approach the deep questions of existence from utterly different starting points. Science is based on careful observation and experiment enabling theories to be constructed which connect different experiences. Regularities in the workings of nature are sought which hopefully reveal the fundamental laws that govern the behaviour of matter and forces. Central to this approach is the willingness of the scientist to abandon a theory if evidence is produced against it. Although individual scientists may cling tenaciously to some cherished idea, the scientific community as a group is always ready to adopt a new approach. There are no shooting wars over scientific principles.

In contrast, religion is founded on revelation and received wisdom. Religious dogma that claims to contain an unalterable Truth can hardly be modified to fit changing ideas. The true believer must stand by his faith whatever the apparent evidence against it. This 'Truth' is said to be communicated directly to the believer, rather than through the filtering and refining process of collective investigation. The trouble about revealed 'Truth' is that it is liable to be wrong, and even if it is right other people require a good reason to share the recipients' belief.

Many scientists are derisory about revealed truth. Indeed, some maintain it is a positive evil:

> Generally the state of mind of a believer in a revelation is the awful arrogance of saying 'I *know*, and those who do not agree with my belief are wrong'. In no other field is such arrogance so widespread, in no other field do people feel so utterly certain of their 'knowledge'. It is to me quite disgusting that anybody should feel so superior, so selected and chosen against all the many who differ in their beliefs or unbeliefs. This would be bad enough, but so many believers do their best to propagate their faith, at the very least to their children but often also to others (and historically there are of course plenty of examples of doing this by force and ruthless brutality). The fact that stares one in the face is that people of the greatest sincerity and of all levels of

> intelligence differ and have always differed in their religious beliefs. Since at most one faith can be true, it follows that human beings are extremely liable to believe firmly and honestly in something untrue in the field of revealed religion. One would have expected this obvious fact to lead to some humility, to some thought that however deep one's faith, one may conceivably be mistaken. Nothing is further from the believer, any believer, than this elementary humility. All in his power (which nowadays in a developed country tends to be confined to his children) must have his faith rammed down their throats. In many cases children are indeed indoctrinated with the disgraceful thought that they belong to the one group with superior knowledge who alone have a private wire to the office of the Almighty, all others being less fortunate than they themselves.[2]

Nevertheless, those who have had religious experiences invariably regard their own personal revelation as a sounder basis for belief than any number of scientific experiments. Indeed, many professional scientists are also deeply religious and apparently have little intellectual difficulty in allowing the two sides of their philosophy to peacefully coexist. The problem is how to translate many disparate religious experiences into a coherent religious world-view. Christian cosmology, for example, has differed radically from Oriental cosmology. At least one must be wrong.

It is a great mistake, however, to infer from the scientist's suspicion of revealed truth that he is necessarily a cold, hard, calculating soulless individual, interested only in facts and figures. Indeed, the rise of the new physics has been accompanied by a tremendous growth of interest concerning the deeper philosophical implications of science. It is a lesser-known side of scientific endeavour, and it frequently comes as a complete surprise. The pathologist, writer and television producer Kit Pedlar describes his astonishment, while planning a television series on mind and the paranormal, at coming across the concern that modern physicists have for broader issues:

> For almost twenty years I occupied my research time as a happy biological reductionist believing that my painstaking research would eventually reveal ultimate truths. Then I began to read the new physics. The experience was shattering.
>
> As a biologist I had imagined the physicists to be cool, clear, unemotional men and women who looked down on nature from a clinical, detached viewpoint — people who reduced a sunset to wavelengths and frequencies, and observers who shredded the

> complex of the universe into rigid and formal elements.
>
> My error was enormous. I began to study the works of people with legendary names: Einstein, Bohr, Schrödinger and Dirac. I found that here were not clinical and detached men, but poetic and religious ones who imagined such unfamiliar immensities as to make what I have referred to as the 'paranormal' almost pedestrian by comparison.[3]

It is ironical that physics, which has led the way for all other sciences, is now moving towards a more accommodating view of mind, while the life sciences, following the path of last century's physics, are trying to abolish mind altogether. The psychologist Harold Morowitz has remarked on this curious reversal:

> What has happened is that biologists, who once postulated a privileged role for the human mind in nature's hierarchy, have been moving relentlessly toward the hard-core materialism that characterized nineteenth-century physics. At the same time, physicists, faced with compelling experimental evidence, have been moving away from strictly mechanical models of the universe to a view that sees the mind as playing an integral role in all physical events. It is as if the two disciplines were on fast-moving trains, going in opposite directions and not noticing what is happening across the tracks.[4]

In the coming chapters we shall see how the new physics has given 'the observer' a central role in the nature of physical reality. A growing number of people believe that recent advances in fundamental science are more likely to reveal the deeper meaning of existence than appeal to traditional religion. In any case, religion cannot afford to ignore these advances.

2. Genesis

'In the beginning God created the heaven and the earth.'
Genesis 1: 1

'But no one was there to see it.'
Steven Weinberg in *The First Three Minutes*

Was there a creation? If so, when did it occur and what caused it? Nothing is more profound or more baffling than the riddle of existence. Most religions have something to say about how things got started; so does modern science. In this book I shall address the enigma of genesis in the light of recent cosmological discoveries. This chapter deals with the origin of the universe as a whole. Some people have used the word 'universe' to mean the solar system or the Milky Way galaxy. I shall use it in the more conventional sense of 'every physical thing that exists', by which I mean all matter distributed among and between all the galaxies, all forms of energy, all non-material things such as black holes and gravity waves, and all of space as well, stretching (if indeed it does) right out to infinity. Sometimes I shall use 'world' to mean the same thing.

Any system of thought that claims to provide an understanding of the physical world must make some statement about the origin of the world. At its most basic, the choice is stark. Either the universe has always existed (in one form or another) or it began, more or less abruptly, at some particular moment in the past. Both alternatives have long been a source of perplexity to theologians, philosophers and scientists, and both present obvious difficulties for the layman.

If the universe had no origin in time — if it has always existed —

then it is of infinite age. The concept of infinity leaves many people reeling. If there has been an infinite number of events already, why do we find ourselves living now? Did the universe remain quiescent for all of eternity only to spring into action relatively recently, or has there been some activity going on for ever and ever? On the other hand, if the universe *began*, that means accepting it appeared suddenly out of nothing. This seems to imply that there was a first event. If so, what caused it? Is such a question even meaningful?

Many thinkers baulk at these issues, and turn instead to the scientific evidence. What can science tell us about the origin of the universe?

These days most cosmologists and astronomers back the theory that there was indeed a creation, about eighteen billion years ago, when the physical universe burst into existence in an awesome explosion popularly known as the 'big bang'. There are many strands of evidence to support this astonishing theory. Whether one accepts all the details or not, the essential hypothesis — that there was some sort of creation — seems, from the scientific point of view, compelling. The reason stems directly from a large body of scientific evidence that is encompassed by the most universal law of physics known — the second law of thermodynamics. In its widest sense this law states that every day the universe becomes more and more disordered. There is a sort of gradual but inexorable descent into chaos. Examples of the second law are found everywhere: buildings fall down, people grow old, mountains and shorelines are eroded, natural resources are depleted.

If all natural activity produces more disorder (measured in some appropriate way) then the world must change *irreversibly*, for to restore the universe to yesterday's condition would mean somehow reducing the disorder to its previous level, which contradicts the second law. Yet at first sight there seem to be many counter-examples of this law. New buildings are erected. New structures grow. Isn't every new-born baby an example of order arising out of disorder?

In these cases you have to be sure you are looking at the total system, not merely the subject of interest. The concentration of order in one region of the universe is always paid for by increasing disorder somewhere else. Take the construction of a new building, for example. The materials used inevitably deplete the world's resources, while the energy expended in the building process is also lost irretrievably. When a full balance sheet is drawn up, disorder always wins.

Physicists have invented a mathematical quantity called entropy to quantify disorder, and many careful experiments verify that the total entropy in a system never decreases. If the system is isolated from its surroundings, any changes that occur within it will remorselessly drive up the entropy until it can go no higher. After that there will be no further change: the system will have reached a condition of thermodynamic equilibrium. A box containing a mixture of chemicals provides a good example. The chemicals will react, some heat may be produced, the constituent substances will alter their molecular form and so on. All these changes increase the entropy inside the box. Eventually, the contents settle down at a uniform temperature in their final chemical form and nothing further happens. To return the contents to their former state is not impossible, but it would mean opening the box and expending energy and materials to reverse the changes that had occurred. This manipulation would produce more than enough entropy to offset the entropy reduction within the box.

If the universe has a finite stock of order, and is changing irreversibly towards disorder — ultimately to thermodynamic equilibrium — two very deep inferences follow immediately. The first is that the universe will eventually die, wallowing, as it were, in its own entropy. This is known among physicists as the 'heat death' of the universe. The second is that the universe cannot have existed for ever, otherwise it would have reached its equilibrium end state an infinite time ago. Conclusion: the universe did not always exist.

We see the second law of thermodynamics at work in all the familiar systems around us. The Earth, for example, cannot have existed for ever, or its core would have cooled down. From radioactivity studies the Earth can be dated to about 4½ billion years, which is similar to the age of the moon and of various meteorites.

As far as the sun is concerned, it clearly cannot continue burning away merrily ad infinitum. Year by year its fuel reserves decline, so that eventually it will cool and dim. By the same token its fires must have been ignited only a finite time ago: it does not have unlimited sources of energy. Estimates place the age of the sun at a little greater than that of the Earth, which accords well with current astronomical theories that the solar system formed together as a single unit. Nevertheless, the solar system is only a minute component of the universe, and it would be rash to draw sweeping conclusions from considerations of the Earth and sun alone. The sun, however, is a typical star, and our galaxy alone contains many billions of other stars whose life cycles can be studied by astronomers. Stars exist that have reached various stages in their evolution, enabling us to build up a detailed picture of stellar birth, life and death.

Stars form, along with planets, as a result of the gradual contraction and fragmentation of huge, tenuous clouds of inter-stellar gas which consist mainly of hydrogen. Today it is easy to find regions of the galaxy where starbirth is taking place. One of these, the Great Nebula in Orion, is visible to the naked eye. The stars were not simply made once and for all. Our sun, for example, at about five billion years old, is at most only half the age of the oldest stars in the galaxy. The formation of the solar system would have been just one further product of a continuing process that has occurred hundreds of billions of times in the Milky Way alone, and will continue in the future. Thus, as far as the formation of stars and planets are concerned, there was no real creation as such at all, merely a sort of cosmic assembly line steadily turning the raw material — hydrogen, helium, and a minute fraction of heavier elements — into stars and planets.

Given that stars are continually burning out while others are being formed to replace them, might this cycle of birth and death have continued endlessly? Alas, no, as the second law of thermodynamics assures us. The material of burnt-out stars can never be fully recycled. The energy needed dissipates away into space in the form of starlight radiated over the aeons. Some of the star stuff is lost irretrievably down black holes.

There is, however, a more direct reason for believing that the entire cosmic system has not been recycling away for all eternity. Isaac Newton, one of the founders of modern science, established that gravity is a universal force, acting between all material bodies in the cosmos: every star, every galaxy, pulls on every other with a gravitational force. Because astronomical bodies float freely in space there seems to be no reason why they do not fall together as a result of this ubiquitous gravitational attraction. In the solar system, gravitational collapse of the planets on to the sun is avoided by centrifugal effects: the planets are revolving around the sun. Likewise the galaxy is rotating. But there is no evidence that the universe as a whole is rotating. Clearly the galaxies can't just hang there for ever. So the universe cannot always have enjoyed its present arrangement.

Although this cosmic conundrum had been appreciated since the time of Newton, it was not until the 1920s that the resolution was discovered. The American astronomer Edwin Hubble found that the galaxies are not falling together because they are rushing apart instead. Hubble noticed that galactic light is slightly distorted in colour ('red shifted' to use the jargon), a circumstance that suggests rapid recession. The reason is that light consists of waves, so a moving light source can stretch or shrink the waves, just as a moving vehicle stretches or shrinks the sound waves it emits. The tone of a car engine, or the whistle of a train, drops dramatically in pitch as it rushes by. In the case of light, read 'colour' for 'pitch' and you have the Hubble red shift. The speeds involved, however, are vastly greater. Distant galaxies recede at many thousands of miles per second.

Hubble's discovery is sometimes misinterpreted to mean that our galaxy is at the centre of this headlong rush, with all the other galaxies flying directly away from us. That is quite wrong. Because the distant galaxies recede faster than the nearby ones, the gaps between the galaxies also expand, so in fact every galaxy is moving away from every other one. This is the famous 'expanding universe'. The pattern of galactic dispersal would appear very much the same from wherever in the cosmos you looked.

The expanding universe accords very well with modern thinking on the nature of space, time and motion. Albert Einstein, who carries the same status in the scientific community as St. Paul does among Christians, revolutionized our conception of these matters with his mind-bending theory of relativity. Although it has taken sixty years for Einstein's spacewarps and timewarps to impinge on the popular imagination, physicists have long accepted his ideas of curved space-time as an explanation of gravity.

The force of gravity powers all large-scale cosmic phenomena. In objects of astronomical size, gravity far outweighs all other forces such as magnetism or electricity. It shapes the galaxies and controls the intergalactic motions. When it comes to explaining the expanding universe, gravity is the key.

Einstein argued convincingly that gravity stretches or distorts space and time, and the idea can be checked directly by watching the sun's gravity bend starbeams that graze its surface. The sky behind the sun appears from Earth to be slightly, but distinctly, bent. The elasticity of time can also be demonstrated, most directly by flying clocks in space. Time runs faster in the gravity-free environment up there than it does on the Earth's surface.

If the sun can stretch space so can the galaxy, which is made of many suns. So rather than thinking of the galaxies as moving apart *through* space, astronomers prefer to think of the space between the galaxies as stretching. If intergalactic space is being 'inflated', then each day every galaxy will find itself with more and more elbow room. In that way the universe expands, without having to expand *into* some external void.

Setting aside for now the concepts of elastic space and time, which many people find hard to understand, it is plainly obvious that a universe which is growing bigger must have been smaller in the past. If the present expansion rate had been maintained throughout history, then twenty or thirty billion years ago the whole observable universe would have been shrivelled up into an unrecognizable blob with no identifiable astronomical bodies at all. In fact, astronomers have discovered that the expansion rate is decellerating somewhat, so this highly compressed condition in fact occurred rather more recently, perhaps fifteen or twenty billion years ago. (Compare the sun's age of five billion years.) Because the expansion rate was much higher then, the early stages of the galactic dispersal resembled an outburst rather than a slow expansion.

It is sometimes said that the universe as we now know it was created by the explosion of a sort of primeval 'egg', the galaxies being fragments of the explosion that are still hurtling through space. It is a picture that captures some correct features but it can also be misleading. The thing that exploded was shrunken because space was shrunken. It is wrong to think in terms of an 'egg' surrounded by a void. An egg has a surface and a middle. Astronomers believe, however, that there is no edge or surface to the universe, and no privileged centre.

We are tangling here with the delicate subject of infinity. It is a topic full of pitfalls for the unwary. In view of its importance not only for the expanding universe, but for the broader issues of science and religion, it is worth a short digression at this stage.

Scientists have long recognized the need to base all their considerations of infinity on precisely formulated mathematical steps, for measuring the infinite can produce all sorts of paradoxes. Consider, for example the famous 'hare and tortoise' paradox due to Zeno of Elea (fifth century B.C.) In a race, the tortoise has a head start, but the hare, running faster, soon overtakes him. Clearly, at every moment of the race the hare is at a place and the tortoise is at a place. As both have been running for the same length of time — for an equal number of moments — then presumably they have passed through an equal number of places. But for the hare to overtake the tortoise he must cover a greater distance in the same time, and so pass through a *greater* number of places than the tortoise. How then can the hare ever overtake the tortoise?

The resolution of this paradox (one of several due to Zeno) involves a proper formulation of the concept of infinity. If time and space are infinitely divisible then both the hare and the tortoise run for an infinity of moments through an infinity of places. The essential feature of infinity here is that a part of infinity is as big as the whole. Although the tortoise's journey is shorter in distance than the hare's, he still covers as many places as the hare (i.e. infinity) — even though we know the hare passes through all the same places as the tortoise, and more!

Many surprises of this sort emerge from a study of the infinite, and it has taken mathematicians centuries of logical construction to fully comprehend the rules for the proper manipulation of infinity. An odd feature is that there exists more than one sort of infinity. There is the infinity of things that can be labelled by whole numbers (1, 2, 3 . . . without end) and a bigger infinity for which even the whole numbers in their entirety are inadequate.

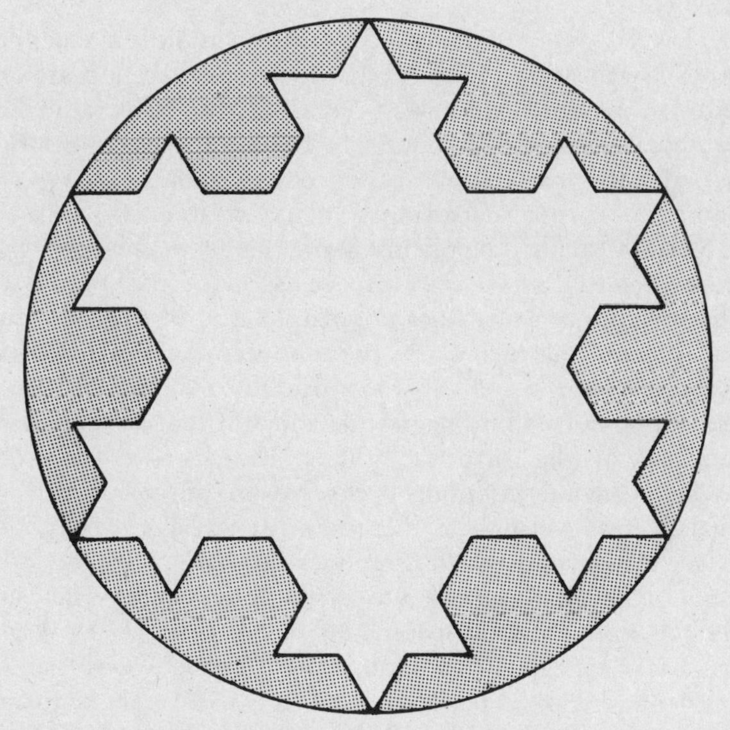

1 The irregular perimeter in this figure is constructed by raising equilateral triangles on the sides of larger triangles in a sequence of steps. The third step is shown in the figure. As the number of steps increases, so the perimeter becomes longer and more 'spikey'. The length of the perimeter grows without limit as the number of steps is increased indefinitely, but the perimeter never protrudes outside the enclosing circle. The area enclosed by the irregular perimeter is therefore finite, even though the length of the perimeter approaches infinity in the limit of an infinite number of steps.

When it comes to geometry, intuition can lead you badly astray. Consider for example the length of a fence that surrounds a field of given area. It is easy to see that a long thin field requires more fence for a given area than a square field. A round field uses the minimum length of fence. But just how long can the perimeter of a field become? Figure 1 shows a rather eccentrically shaped perimeter consisting of triangles built upon triangles in a sequence of steps. With each step the perimeter fence gets longer, and the area enclosed increases a bit. But the perimeter will never protrude beyond the enclosing circle, so the area will always remain finite, yet the perimeter can grow without limit as the number of additional triangular wedges is increased. It is thus possible to conceive of an *infinitely* long fence enclosing a *finite* area of field (see Fig. 1).

What has all this got to do with the creation of the universe? First, it illustrates that ideas like 'infinity' should not be used sloppily or they are likely to produce nonsense. Secondly, it demonstrates that the results obtained often run counter to common sense and intuition. This is one of the great lessons of science. It is often necessary to resort to the abstract — to formal mathematical manipulations — to make sense of the world. Ordinary experience alone can be an unreliable guide.

Is the universe infinite in size? If space has an infinite volume we can envisage an infinity of galaxies populating it with roughly uniform density. Many people then worry about how something that is infinite can expand. What is there for it to expand into? There is no problem: infinity can be boosted in magnitude and still remain the same size. (Remember what the 'tortoise taught us'.) But visualization problems set in when we wind this model backwards to the 'cosmic egg' phase. If the galaxies are everywhere, there could never have been a *finite* egg, with a surface beyond which there was no matter. So eggs are out.

Imagine, in such an infinite universe, a huge sphere enclosing an enormous volume of space containing many galaxies. Now picture space everywhere rapidly shrinking, like Alice in Wonderland after eating the magic cake. The sphere contracts to a smaller and smaller radius; but however shrunken it becomes there is still unending space and an infinity of galaxies outside it. If the sphere shrinks to literally nothing, then we have the mathematically delicate problem of an infinite universe which is infinitely shrunken. There is still no centre or edge, but the contents of any sphere, however large it started out, would be crushed together into a single point. Astronomers believe that it was from such an infinitely shrunken, yet unbounded, state that the universe exploded.

There is, in fact, another possible model for the universe that avoids the competition of infinities; it was proposed by Einstein himself in 1917. Based on the fact that space can bend, Einstein argued that space can connect up to itself in a variety of unexpected ways. The curved surface of the Earth can be used as an analogy. The Earth's surface is finite in area, but unbounded: nowhere does a traveller meet an edge or boundary. Similarly space could be finite in volume, but without any edge or boundary. Few people can really envisage such a monstrosity, but mathematics can take care of the details for us. The shape is called a hypersphere. If the universe is a hypersphere an astronaut could, in principle, circumnavigate it like a cosmic Magellan by always pointing his rocket in the same direction until he returned to his starting point.

Although it is finite, Einstein's hyperspherical cosmos still has no centre or edge (just as the surface of the Earth has no centre or edge), so when shrunken it does not resemble a cosmic egg either. One can imagine the hypersphere shrivelling away to nothing, its volume vanishing, analogous to the surface of a sphere being shrunk to zero radius (see Fig. 2).

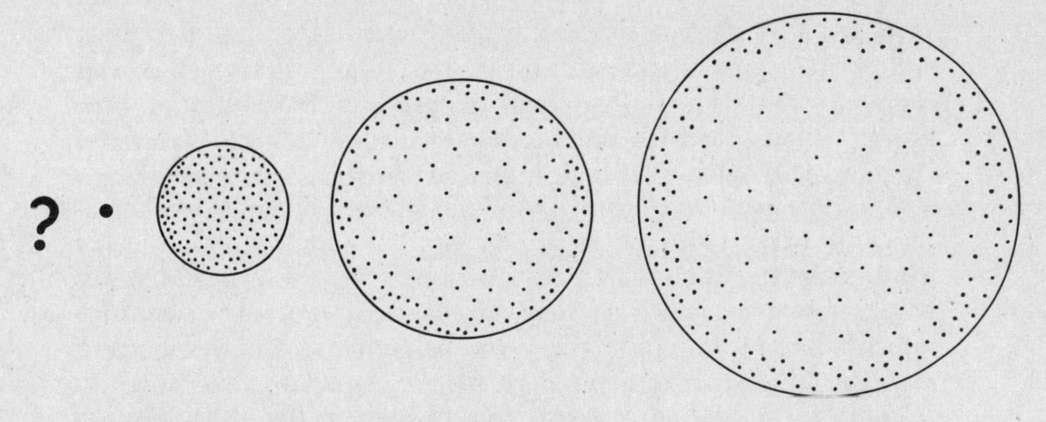

2 If three-dimensional space is represented by a two-dimensional surface, then one model of the expanding universe is reminiscent of a balloon that inflates from nothing. In this model space is finite, but unbounded: an observer in the space could travel freely all around the universe. The dots represent galaxies (or clusters of galaxies). As the universe expands, space stretches, so all the dots move farther apart from all their neighbours. An observer on any one of the dots would see the other dots receding in a systematic pattern, and would seem to be at the centre of this outward migration.

The study of elastic space has led cosmologists to propose a theory of the creation which differs greatly in detail from the biblical version.

The most startling feature of the scientific theory is the suggestion that space itself was created in the big bang, and not merely matter. If the 'shrivelling balloon' model is envisaged instead as an expanding balloon — expanding out of nothing — then you obtain a rough idea of the story of Genesis as told by modern physics. The important point is that continuation of the concept of space back through the infinitely shrunken phase is impossible, and this is true whether or not the universe resembles Einstein's hypersphere (balloon model) or is infinite in size. The first instant of the big bang, where space was infinitely shrunken, represents a boundary or edge in time at which space ceases to exist. Physicists call such a boundacy a *singularity*.

The idea of space being created out of nothing is a subtle one that many people find hard to understand, especially if they are used to thinking of space as already being 'nothing'. The physicist, however, regards space as more like an elastic medium than as emptiness. Indeed, we shall see in later chapters that, because of quantum effects, even the purest vacuum is a ferment of activity and is crowded with evanescent structures. To the physicist 'nothing' means 'no space' as well as no matter.

More peculiarities lie in wait. Space is inextricably linked to time, and as space stretches and shrinks, so does time. Just as the big bang represents the creation of space, so it represents the creation of time. Neither space nor time can be extended back through the initial singularity. Crudely speaking, time itself began at the big bang.

These bizarre ideas can only be fully grasped by appeal to mathematics. Human intuition is an inadequate guide — which illustrates one of the principal reasons for the success of the scientific method. By employing mathematics as a language, science can describe situations which which are completely beyond the power of human beings to imagine. Indeed, most of modern physics falls into this category. Without the abstract description provided by mathematics, physics would never have progressed beyond simple mechanics. Of course, physicists, like everybody else, carry around mental models of atoms, light waves, the expanding universe, electrons, and so on, but the images are often wildly inaccurate or misleading. In fact, it may be logically impossible for anyone to be able to correctly visualize certain physical systems, such as atoms, because they contain features that simply do not exist in the world of our experience (as we shall see when we come on to look at the quantum theory in Chapter 8).

Failure of the human imagination to grasp certain crucial features of reality is a warning that we cannot expect to base great religious truths (such as the nature of the creation) on simple-minded ideas of space, time and matter, gleaned from daily experience.

Intellectual difficulties over the origin of time are not new. Aristotle and St. Thomas Aquinas both rejected the idea of time being created, for that would imply there was a first event. What caused the first event? Nothing, for there was no prior event.

The finitude of time, in fact, need not imply that there was a first event. Imagine events labelled by numbers, with zero corresponding to the singularity. The singularity is not an event, it is a state of infinite density, or something like it, where spacetime has ceased. If one now asks, 'What is the first event *after* the singularity?', this is the same as the question, 'What is the smallest number greater than zero?' There is no such number, for every fraction, however small, can always be halved. Likewise, there is no first event.

The trouble is that infinite time is equally perplexing, as Immanuel Kant later emphasized:

> If we assume the world has no beginning in time, then up to every given moment an eternity has elapsed, and there has passed away in the world an infinite series of successive states of things. Now the infinity of the series consists of the fact that it can never be completed through successive synthesis. It thus follows that it is impossible for an infinite world series to have passed away, and that a beginning of the world is therefore a necessary condition of the world's existence.[1]

Remembering Zeno, however, we must be wary of manipulating infinity. According to Kant's reasoning, the hare could never complete the infinite series of steps 'through successive synthesis' necessary for him to overtake the tortoise. Yet we all know that he will. Nor is it a valid objection to point out that in the Zeno case the elapsed time is finite, while Kant refers to the passage of an infinite duration. In both cases there is an infinity of moments involved. Any mathematician can demonstrate that there are no more moments in all of eternity than there are in, say, one minute. In both cases there is an infinite number, and this infinity can be made no bigger by 'infinite stretching'.

Another objection to Kant's reasoning is the assumption that time 'elapses', which implies a flowing or moving time. Few physicists would concede that time does flow or move. It is simply *there*, like space (a topic that we shall return to in Chapter 9).

In conclusion, there seems to be nothing terribly wrong with either an eternal universe, or one that is of a finite age, bounded in the past by a singularity. Assuming the latter to be correct, does this mean that science supports the biblical version of the creation?

There is no agreement among Christians on the weight to be placed on the biblical narrative of Genesis. In 1951, Pope Pius XII, addressing the Pontifical Academy of Sciences in Rome on the implications of modern scientific cosmology,[2] alluded to the big bang theory, and the fact that 'everything seems to indicate that the universe has in finite times a mighty beginning'. His remarks provoked a fierce reaction (not least among scientists), and contemporary theologians are still divided over whether the big bang is the creation event supposedly revealed to the Bible writers. Thus, Ernan McMullin of Notre Dame University in the United States, writing recently under the title, 'How should cosmology relate to theology?', concludes that: 'What one cannot say is, first, that the Christian doctrine of creation "supports" the Big Bang model, or second, that the Big Bang model "supports" the doctrine of creation.'[3] Nevertheless, many laymen, compelled these days to dismiss so much of the Old Testament as fiction, find comfort in the apparent support that modern scientific cosmology brings to the Genesis story.

If we accept that space and time really did erupt out of nothing in the big bang, then clearly there was a creation and the universe has a finite age. The paradox of the second law of thermodynamics is therefore immediately solved. The universe has not reached thermodynamic equilibrium yet because it has only been disordering itself for eighteen billion years or so, and that is nowhere near long enough to complete the process. Moreover, we can now understand why all the galaxies have not fallen together. The explosive violence has flung them apart, and even though their rate of separation is slowing, there has not yet been enough time for them to fall back on themselves.

If the big bang theory rested on the work of Hubble and Einstein alone, it would not command the widespread support that it does. Fortunately, there is some persuasive confirmatory evidence.

The searing violence which accompanied the birth of the cosmos must have left many imprints on the structure of the universe, and we might expect some relics of the primeval phase to survive today. Searching for relics from the creation is now one of the most popular scientific enterprises and, incredible though it may seem, there are good financial reasons for this. The primeval universe provided an ideal natural laboratory in which physical conditions of such extremity were realized that they cannot be simulated on Earth with even the most elaborate scientific equipment. To test their theories about the behaviour of matter under these extreme conditions physicists must appeal to the cosmology of the newly-created universe. The hope is that the universe today may contain traces or clues about the physical processes that occurred during that first brief flash of existence. Calculations may then be used to see if those processes accord with what the theorists expect for the behaviour of matter under extreme conditions.

By far the most important relic of the primeval universe was discovered by accident in the mid-1960s. Two physicists working for the Bell Telephone Company stumbled across some mysterious radiation coming from space. A careful analysis has revealed that this radiation, which bathes the whole universe, is a relic of the primeval heat, the last fading glow of the fiery birth of the universe. The big bang, like any explosion, generated huge quantities of heat. Indeed, it took 100,000 years for the cosmic gases to cool to the sort of temperatures found at the surface of the sun. Now, eighteen billion years on, the temperature has dropped to the very depths, a mere three degrees above absolute zero (−273°C). Nevertheless there is still a vast amount of energy locked up in the heat radiation.

Knowing the present temperature of the relic heat radiation, it is a simple matter of scaling to compute its value at all epochs. Every time a typical region of the universe doubles in size, the temperature falls by fifty per cent. Working backwards, it is readily deduced that, for example, at one second after the creation, the temperature was ten billion degrees. This may seem pretty hot, but it is well within the range of laboratory experience. Indeed, using modern particle accelerators to generate high energy collisions, it is possible to simulate for a fleeting instant the conditions in the primeval explosion at a mere million-millionth of a second after the beginning, when the temperature was a staggering million billion degrees. It is therefore with some confidence that astrophysicists can model many of the physical processes that must have occurred after that first searing instant.

Using such models it is possible to compute the form of the cosmic material at each epoch as the universe erupted into existence. For example, between about one second and five minutes conditions would have been suitable for nuclear reactions to have occurred. The major process would have been the fusion of hydrogen nuclei to form helium and some deuterium. Calculations predict that the final ratio of helium to hydrogen should be about twenty-five per cent by mass, which is very close to what is observed to be the relative cosmic abundances of these two elements today. (Hydrogen and helium together constitute over ninety-nine per cent of the material in the universe.) Such remarkable agreement gives us confidence that the basic ideas of the hot big bang theory are correct.

The epochs before one second, being so hot, involved some very high energy physics. At these temperatures matter is broken completely apart, and its primary constituents (to be discussed in Chapter 11) would have been exposed. This very early phase — the first one second of existence — is now the subject of intense study by theoretical physicists, some of whom believe that many of the features of the universe can be explained by processes that occurred then. In the next chapter some of these more recent developments will be described.

The big bang theory is now taken very much for granted by astrophysicists, and the helium abundance calculations have long become part of standard cosmology. It is therefore easy to overlook the remarkable nature of these early successes. Had a nineteenth-century archaeologist claimed to have discovered the Garden of Eden and produced a relic showing unmistakable evidence of God's handiwork during the first day, the claim would have produced a sensation. Helium may not be very familiar to most people, but it can readily be purchased from industry. It is an extraordinary thought that this commonplace laboratory substance was fashioned in the primeval furnace, not just during the first day, but in the first few minutes of existence.

Though present scientific opinion lends strong support to the creation theory, it is important to realize that there is no *logical* reason why the universe cannot be infinitely old. The chief physical difficulty is, as we have seen, the second law of thermodynamics. However, from time to time mechanisms have been proposed to overcome this difficulty. One of these is the steady-state theory, due to Hermann Bondi, Thomas Gold and Fred Hoyle. In all versions of this theory the universe is infinite in age, but the thermodynamic heat death is avoided by postulating that new low-entropy matter is continually being created. Thus, rather than matter appearing all in one go in a primeval explosion, it arises gradually, or perhaps sporadically in mini-bangs, over the aeons. The average rate of appearance of new matter is adjusted (perhaps by a feedback mechanism) so that, as the universe expands and the density of existing matter is diluted, the newly-created matter fills in the gaps and maintains a roughly constant density. The dispersal of the galaxies is thereby compensated by the creation of new galaxies in the widening void in such a way that the overall aspect of the universe remains much the same from epoch to epoch. Globally, nothing changes (see Fig. 3). In contrast, in the big bang model, the density of galaxies steadily declines, and the universe

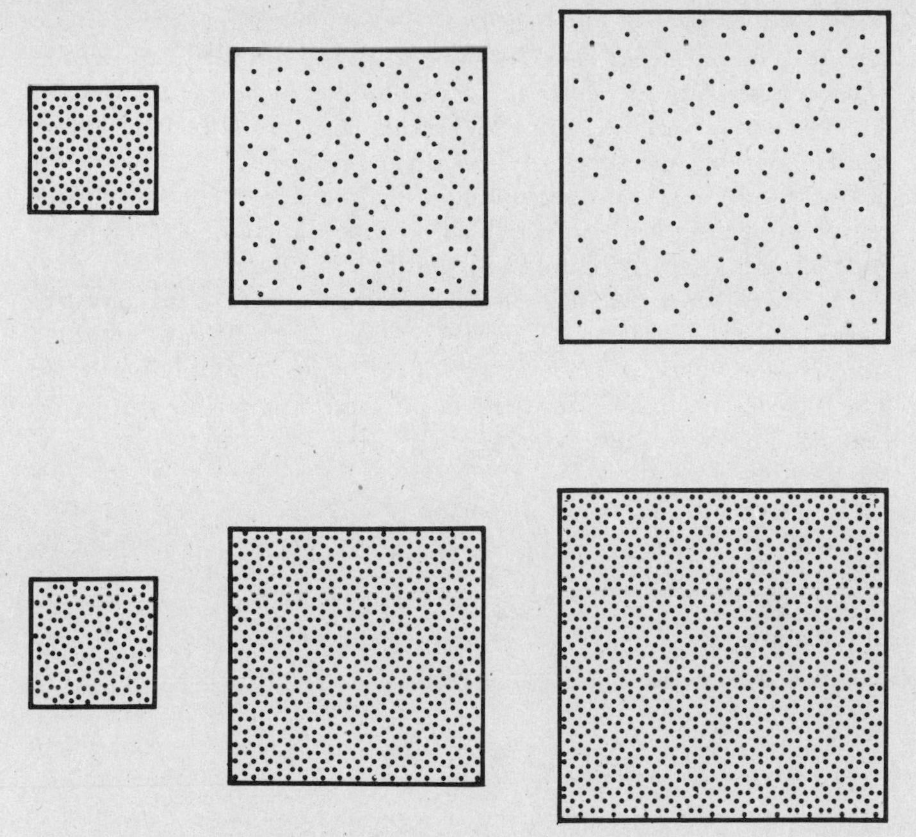

3 The figure contrasts three successive 'snapshots' of a region of expanding space in the big bang and steady-state models of the universe. In the big bang case (upper) the number of galaxies (dots) remain unchanged in a given volume of space. Thus, the density of dots declines as the expansion proceeds. In the steady-state case (lower) the density of galaxies remains unchanged from epoch to epoch, so that new galaxies must be continuously created to fill in the gaps made by the expanding space.

evolves in structure and arrangement.

Hoyle attempted to explain the continual creation of matter by inventing a new type of field that carries negative energy. The steady enhancement of this field pays for the positive energy necessary to create the matter. (The creation of matter from energy will be described in the next chapter.) Thus, God is abolished from the steady-state model altogether. First, the primary energy necessary to create matter need not be created; it is simply paid for by depositing *negative* energy into some other system. Secondly, space and time are not created, but have always existed.

The steady-state model had great philosophical appeal for many scientists, who were attracted by its elegance and simplicity. However, advances in astronomy put paid to any simple version of the theory, and the discovery of the cosmic background heat radiation in 1965 was really the last nail in its coffin. It remains an important idea, though, for it demonstrates the logical possibility of a universe with neither abrupt creation nor heat death, in which all processes, including the appearance of matter, are attributed to natural mechanisms.

The fact that modern cosmology has provided hard physical evidence for the creation is a matter of great satisfaction to religious thinkers. However, it is not enough that a creation simply occurred. The Bible tells us that God created the universe. Can science throw any light at all on what caused the big bang? This will form the subject of the next chapter.

3. Did God create the universe?

'I want to know how God created this world.'

Einstein

'I had no need of this hypothesis.'

Pierre Laplace to Napoleon Bonaparte

A well-known periodical recently proclaimed in banner headlines 'Astronomers discover God!'. The subject of the article was the big bang and recent advances in our understanding of the very early epochs of the universe. In the world of popular journalism, the fact of creation is itself considered sufficient to reveal the existence of God. But what does it really mean to say that God *caused* the creation? Is it possible to conceive of a creation without God? Does modern astronomy inevitably expose the limits of the physical universe and compel us to invoke the supernatural?

The word 'creation' carries with it a variety of meanings, and it is important to keep a clear distinction between them. The creation of the universe can be taken to mean the abrupt organization of matter from a chaotic, structureless primeval form into the currently observed complex order and elaborate activity. It can mean the actual creation of matter in what was previously a featureless void. Or it can mean the abrupt appearance of the entire physical world, including space and time, from nothing at all. There is also the separate issue of the creation of life and man himself, which we shall deal with later.

The biblical version of the creation of the universe 'on the first day' is vague about exactly what was involved. There are actually two accounts of the creation, but neither explicitly mentions that the material from which the stars and planets, the Earth, and our own bodies are made, existed prior to the creation event. The belief that God created this cosmic material out of nothing is a longstanding part of Christian doctrine. Indeed, it seems to be demanded by the assumption of God's omnipotence, for if God did not create matter, it would imply that he was limited in his work by the nature of the raw materials available to him.

Before this century it was assumed by scientists and theologians alike that matter could not be created (or destroyed) by natural means. Of course, the form of matter can change, for example, during chemical reactions, but the total quantity of matter was considered to be, without exception, constant. Scientists, faced with the problem of the origin of matter, were inclined to believe in a universe of infinite age, thereby avoiding the need for a creation altogether. In an eternal universe, matter can have existed for ever, and the problem of its origin is side-stepped.

The belief that matter cannot be created by natural means collapsed dramatically in the 1930s when matter was first made in the laboratory. The events leading up to this discovery provide a classic example of modern physics at its best.

This story, as with so many others, began with Einstein in 1905. His famous $E = mc^2$ equation is the mathematical embodiment of the statement that mass and energy are equivalent: mass has energy and energy has mass. Mass is the quantification of matter: the mass of a body tells you how much matter it contains. Large mass means heavy and ponderous, small mass means light and easy to move. The fact that mass is equivalent to energy means that, in a sense, matter is 'locked up' energy. If some way can be found to unlock it, matter will disappear amid a burst of energy. Conversely, if enough energy is somehow concentrated, matter will appear.

In its original conception, Einstein's equation, a by-product of his theory of relativity, was concerned with the properties of bodies moving at ultra-high speed, close to the speed of light. According to the theory, the energy of the body's motion ought to result in it appearing to be heavier (to increase in mass). The effect is minute at ordinary speeds because a little mass is worth an awful lot of energy: for example, one gram is the equivalent of a million dollars in energy at current prices. However, modern subatomic particle accelerators can boost the speed of electrons and protons to within a whisker of the speed of light, where their masses are observed to increase dozens of times.

Increase of mass with speed does not, of course, amount to the actual creation of matter. Rather, it involves already existing matter putting on weight. The possibility of completely new particles of matter being produced out of concentrated energy emerged with the epoch-making mathematical investigations of Paul Dirac about 1930. Dirac was attempting to reconcile Einstein's theory of relativity, with its $E = mc^2$, with the other major revolution in twentieth-century physics, the quantum theory, concerned with the behaviour of atomic and sub-atomic matter. A unified relativistic quantum theory is needed to describe subatomic particles moving at near the speed of light, such as occurs as a result of energetic radioactive emissions.

Following a mathematical analysis, Dirac suggested a new equation to describe high speed atomic matter. It was an immediate success because it explained a hitherto baffling property known to be possessed by electrons, namely, that they are spinning in a fashion totally at odds with either commonsense or elementary geometry. Crudely speaking, an electron has to turn around twice before it presents the same face as before. It provides another good example of how mathematics must replace intuition in the abstract world of fundamental physics.

Dirac's equation did, however, have one puzzling aspect. Its solutions correctly described the behaviour of ordinary electrons, but for every such solution there existed another associated solution which did not appear to correspond to anything in the known universe. With a bit of imagination it was possible to work out what these unknown particles would be like. In mass and spin they would be identical to ordinary electrons, but whereas all electrons carry negative electric charge, the new mystery particles would have positive charge. Other properties, such as their spin, would also be reversed, making the new particles a sort of mirror image of electrons.

More spectacular was Dirac's prediction that if enough energy could be concentrated, one of these 'antielectrons' might appear where none had existed before. In order that electric charge be conserved, this event would have to be accompanied by the simultaneous appearance of an electron too. In this way energy might be directly used to create matter in the form of an electron–antielectron pair.

About this time (1930), the physicist C.Y. Chao had been experimenting with the penetrating power of gamma rays (high energy photons of light) in heavy materials such as lead. He noticed that the most energetic gamma rays were being attenuated in a curiously efficient manner. The cause of the additional absorption of the rays was a mystery to Chao, but we now know it was caused by electron–antielectron pair production.

Then, in 1933, Carl Anderson was studying the absorption of cosmic rays — high energy particles from space — by metal sheets, when he spotted, for the first time, the unambiguous appearance of Dirac's antielectron. Matter had been created in the laboratory in a controlled experiment. It was quickly verified that the new particles possessed all the properties expected of it, and Dirac and Anderson shared a Nobel prize for this brilliant prediction and discovery.

In subsequent years, the production of electrons and antielectrons (usually called positrons) became commonplace in a wide range of laboratory processes. After World War II, the development of subatomic particle accelerating machines enabled the controlled production of other types of particles too. Antiprotons and antineutrons were made. Today, positrons and antiprotons can be made in large quantities and stored in magnetic 'bottles'. Collectively, the mirror or antiparticles are known as antimatter, and it is now made routinely in physics laboratories.

Armed with these facts the way seems open for a natural explanation of the origin of all matter. During the big bang, huge quantities of energy were available to cause the incoherent production of vast amounts of matter and antimatter. Eventually, much cooled, this material would have aggregated into stars and planets. Unfortunately there is a major snag with this simple idea. When antimatter encounters matter, the two annihilate each other with a violent release of energy — the reverse process of matter creation (see Fig. 4).

A universe consisting of a mixture of matter and antimatter is therefore violently unstable. Very stringent limits can be placed on the admixture of antimatter in our galaxy, and it is a trifling amount. So where has all the antimatter gone? In the laboratory, every particle that is created is accompanied by an antiparticle, so we might expect the universe to be a fifty–fifty mix, but the observations rule this out. Some astrophysicists have attempted to explain this enigma by hypothesizing that somehow matter and antimatter managed to separate into large domains consisting predominantly of one type or the other. Perhaps whole galaxies are made of antimatter and others of matter. However, no convincing mechanism for separating matter from antimatter has ever been proposed, and the symmetric-universe theory has fallen into disfavour.

Those scientists who insisted that the big bang was the creation were thereby faced with the apparent necessity of assuming that some

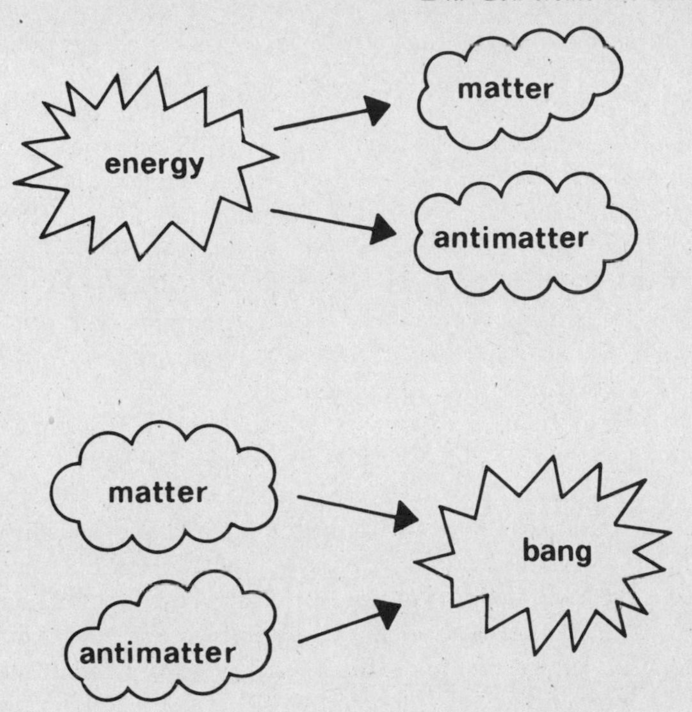

4 In the laboratory, energy can be used to create matter, but it is always accompanied by an equal quantity of antimatter. When matter and antimatter meet, explosive annihilation results, releasing the energy encapsulated in the material. There is a mystery in how all the matter in the universe was created without polluting it with a dangerous mixture of antimatter.

supernatural process had injected matter into the universe without antimatter, in defiance of the laws of physics. Vague excuses about 'all laws break down at the singularity anyway' did little to lessen the feeling of unease.

Very recently, however, a possible route out of this dilemma has appeared. Although under laboratory conditions the creation of matter and antimatter is always symmetric, in the ultra-high temperatures of the big bang it is possible that a very slight excess of matter was permitted. The idea stems from a programme of theoretical work that seeks to provide a unified description of nature's four fundamental forces (a topic to be discussed more fully in Chapter 11). According to the theoretical calculations, at a temperature of a billion billion billion degrees, which could have been attained only during the first billion-billion-billion-billionth of a second, for every billion antiprotons, one-billion-and-one protons were created. Similarly, electrons would have outnumbered positrons by one part in a billion.

Such an excess, while minute, would be crucially significant. In the subsequent carnage, the billion matched pairs of protons and antiprotons would have annihilated each other, but the single unpaired proton would have survived, along with a solitary electron. These left-over particles — almost an afterthought of nature — became the material that eventually formed all the galaxies, all the stars and planets — and us. According to this theory, our universe is built out of a tiny residue of unbalanced matter that survives as a relic of the first unthinkably brief moment of existence.

Like all good theories, physicists find this explanation for the origin of matter persuasive. But where is the hard evidence?

Two confirmatory results seem possible. The first concerns the wholesale annihilation of the billion matched particle–antiparticle pairs that accompanied each excess particle at the outset. The energy of this slaughter must also survive, presumably in the form of heat. But as mentioned in the previous chapter, the universe is indeed bathed in heat radiation left over from the big bang. It is a simple matter to tot up the heat energy per surviving atom to see if the numbers square up with the one-in-a-billion calculation. They do; or at least agreement can be achieved with very plausible models. So not only is the origin of matter explained by this theory, but also the precise temperature of the universe. It is a remarkable achievement.

Nevertheless, some further confirmation is desirable before one can confidently pronounce that matter no longer requires a divine origin. Some form of direct laboratory evidence of a distinct asymmetry between matter and antimatter would carry the greatest conviction. With luck, we may be on the threshold of obtaining just that sort of evidence.

The theory that predicts the tiny excess of matter production also predicts a minute spontaneous *destruction* of matter by the same mechanism. Over an immense duration of time protons, so the theory goes, will decay into positrons, which will then go on to annihilate electrons. In this way all matter is destined ultimately to disappear. The time scale is, however, so long that the human body, on average, looses only about one of its protons per lifetime. To test this theory, scientists are studying huge accumulations of matter, well below ground to cut out the polluting effect of cosmic rays, to try and catch a disappearing proton in the act. Because the process is statistical in origin (like radioactivity), the occasional freak decay will be observed after a modest wait of several weeks, even though the average lifetime of a proton is at least ten thousand billion billion billion years. The secret is to amass many tons of material (which represents a lot of protons) to spot the occasional random event. Several such experiments are currently in progress, and at least one has produced some possible proton-decay events.

The question of the origin of matter illustrates a fundamental problem that faces any attempt to deduce the existence of God from physical phenomena. What once seemed miraculous — the appearance of matter without antimatter — perhaps requiring a supernatural input at the big bang, now seems explicable on ordinary physical grounds, in the light of improved scientific understanding. However astonishing and inexplicable a particular occurrence may be, we can never be absolutely sure that at some distant time in the future a natural phenomenon will not be discovered to explain it.

Do these scientific advances mean that we can now explain the creation in terms of natural processes? Many theologians would strongly deny this. The processes described here do not represent the creation of matter out of nothing, but the conversion of pre-existing energy into material form. We still have to account for where the energy came from in the first place. This surely requires a supernatural explanation?

Nevertheless, one must be careful about shifting the responsibility from matter to energy this way. Energy is a rather slippery concept, especially in modern physics. What is energy? It can take many different forms. It might simply be motion, for example. In the laboratory, particles can collide at high speed and four appear where previously there were only two. The newcomers are paid for by reducing the speed of the two original particles. The conversion of motion, which is intangible, into stuff, which can be kicked, comes very close to the spirit of creation out of nothing.

There is a still more remarkable possibility, which is the creation of matter from a state of *zero* energy. This possibility arises because energy can be both positive and negative. The energy of motion or the energy of mass is always positive, but the energy of attraction, such as that due to certain types of gravitational or electromagnetic field, is negative. Circumstances can arise in which the positive energy that goes to make up the mass of newly-created particles of matter is exactly offset by the negative energy of gravity or electromagnetism. For example, in the vicinity of an atomic nucleus the electric field is intense. If a nucleus containing 200 protons could be made (possible but difficult), then the system becomes unstable against the spontaneous production of electron–positron pairs, without any energy input at all. The reason is that the negative electric energy generated by the new pair of particles can exactly offset the energy of their masses.

In the gravitational case the situation is still more bizarre, for the gravitational field is only a spacewarp — curved space. The energy locked up in a spacewarp can be converted into particles of matter and antimatter. This occurs, for example, near a black hole, and was probably also the most important source of particles in the big bang. Thus, matter appears spontaneously out of empty space. The question then arises, did the primeval bang possess energy, or is the entire universe a state of zero energy, with the energy of all the material offset by negative energy of gravitational attraction?

It is possible to settle the issue by a simple calculation. Astronomers can measure the masses of galaxies, their average separation, and their speeds of recession. Putting these numbers into a formula yields a quantity which some physicists have interpreted as the total energy of the universe. The answer does indeed come out to be zero within the observational accuracy. The reason for this distinctive result has long been a source of puzzlement to cosmologists. Some have suggested that there is a deep cosmic principle at work which requires the universe to have exactly zero energy. If that is so the cosmos can follow the path of least resistence, coming into existence without requiring any input of matter or energy at all.

Matters are further complicated by the fact that energy is not even properly defined when gravity is present. In some cases it is possible to make sense of the total energy in an isolated system by considering its gravitational influence a great (in fact infinite) distance away. But this strategy fails completely in the case of a universe that is spatially finite, such as the model proposed by Einstein and discussed briefly in the previous chapter. In such a closed universe, the total energy is a meaningless quantity.

Do these examples, such as the natural creation of matter out of empty space, perhaps with no need for even an energy input, amount to the creation *ex nihilo* of theology? It could be argued that science has still not explained the existence of space (and time). Granted that the creation of matter, for so long considered the result of divine action, can now (perhaps) be understood in ordinary scientific terms, is it only by an appeal to God that one can explain why there is a universe at all — why space and time exist in the first place, that matter may emerge from them?

The belief that the universe as a whole must have a cause, that cause being God, was enunciated by Plato and Aristotle, developed by Thomas Aquinas, and reached its most cogent form with Gottfried Wilhelm von Leibniz and Samuel Clarke in the eighteenth century. It is usually known as the cosmological argument for the existence of God. There are two versions of the cosmological argument: the causal argument, to be considered here, and the argument from contingency which will be discussed in the next chapter. The cosmological argument was treated with scepticism by David Hume and Immanuel Kant and has been bitterly attacked by Bertrand Russell.

The goal of the cosmological argument is two-fold. The first is to establish the existence of a 'prime mover' — a being that in turn accounts for the existence of the world. The second is to prove that this being is indeed the God as usually understood in Christian doctrine.

The argument proceeds along the following lines. Every event, it is maintained, requires a cause. There cannot be an infinite chain of causes, so there must be a first cause of everything. This cause is God. Now it must be stated right at the outset that there have been many versions of the cosmological argument, and many subtle interpretations of meaning, so that over the years the debate has become rather esoteric and complex. I make no attempt here to give a balanced appraisal of the pros and cons, save only to say that the argument has engaged the attention of some of the greatest intellects in the history of mankind, which has nevertheless not prevented both proponents and opponents of the thesis from making logical and philosophical blunders. Our concern here is to re-examine the causal chain hypothesis in the light of modern science.

Let us examine the first step in the argument: every event has a cause. As Clarke declared: 'Nothing can be more absurd, than to suppose that anything is; and yet that there be absolutely no reason why it is, rather than not.'[1] One usually assumes, loosely speaking, that everything that happens is caused to happen by something else and every object that has come into existence has been produced by something already existing. It seems reasonable enough, but is it true?

In daily life we rarely doubt that all events are caused in some way. A bridge falls down because it is overloaded, snow melts because the air warms up, a tree grows because a seed has been planted, and so on. But do some things have no cause?

Consider the above assertion 'every object that has come into existence has been produced by something'. What if an object has never come into existence, but has always existed? Such a thing is certainly conceivable: space in the steady-state universe, for example. Does it mean anything to ask whether an eternally existing object — one which at no time did not exist — has a cause? One could still ask 'Why does it exist?' The retort 'It has always done so' seems rather lame. Since one can well imagine that the object might not exist, it seems legitimate to seek a reason why it exists rather than not, irrespective of its infinite age. So, in the opinion of some, abolishing the creation (as in the steady-state universe) does not remove the necessity of explaining why a universe exists at all.

Turning aside from the issue of eternal objects for the moment, suppose we restrict ourselves to the coming-into-being of objects. Can something be created out of nothing? We saw how particles can be created out of empty space, but in that case the spacewarp was the cause. We still have to explain where space came from (if it hasn't always existed). Some people might question whether space is a *thing*. Certainly it is hard to imagine Thomas Aquinas or Leibniz regarding it as part of the causal chain. Still, let us press on. What caused space to suddenly appear in the big bang? The singularity? But a singularity is most certainly not a thing. It is the boundary of a thing (spacetime). Impasse.

Does every *event* have a cause? Can something happen without any prior action, or any rational reason? Newspapers often proclaim 'Object in sky unexplained'. This does not mean, however, that aerial phenomena occur that have *no* explanation, only that there is no *known* explanation. Unfortunately it is hard to see how the assertion 'every event has a cause' could ever be definitively falsified, for to do so one would not only have to find an event for which no cause is apparent, but go on to demonstrate that however much information one had about the universe and however deep one's understanding of nature, no cause would ever be found. That seems to be impossible. How can one be sure that the event in question is not caused by some totally obscure, exceedingly rare, never-before-encountered, unobtrusive, freak process?

The nearest science has come to falsifying the claim that every event has a cause is quantum mechanics. As we shall see in Chapter 8, in the subatomic world the behaviour of particles is generally unpredictable. From one moment to the next you cannot be sure what a particle is going to do. If for an event one were to choose the arrival of a subatomic particle at a particular place then, according to the quantum theory, that event has no cause, in the sense that it is inherently unpredictable. No matter how much information is available about the forces and influences acting on the particle, there is no way that its arrival at the designated place can be regarded as 'fixed' by anything else. The outcome is intrinsically random. The particle just pops up in that place with no rhyme nor reason.

Some (a minority) of physicists have not taken kindly to this idea. Einstein dismissed it in a famous retort: 'God does not play dice.' These physicists desire that every event should be caused by something or other, even at the subatomic level. Amazingly enough, it is possible to perform an experiment to demonstrate that, unless influences can travel faster than light, atomic systems are indeed inherently unpredictable — 'God' *does* play dice. Subject to the proviso that some extraordinary conspiracy of nature has not confounded the experimental results, this claim seems to be on a fairly firm foundation.

If, therefore, one accepts that quantum events have, individually, no direct cause, then can the creation of matter, which is a classic example of a quantum process, be said to be without physical cause? In a sense, yes. An *individual* particle will come into existence abruptly and unpredictably, at no specially designated place or moment. However, its behaviour, while maverick, is still subject to the laws of probability. Given a spacewarp of a particular strength, it can be very probable that a particle will appear in a given volume of space in a certain interval of time. But never definite. Conversely, though the probability is exceedingly small, there is still a finite chance of such a particle popping out of nowhere in your living room right now. In the quantum world, such things happen without warning. The fact that the *probability* of particle creation depends on the strength of the spacewarp implies a sort of loose causation. The spacewarp makes the appearance of a particle *more likely*. Whether that is to be regarded as strictly the *cause* of the particle's appearance is largely a matter of semantics.

Now it might be objected that the central discussion concerns whether or not the entire universe has a cause, not whether an electron's creation or arrival at a place has a cause. Some physicists would doubtless reply that the whole universe is also subject to quantum principles, but this is to take us into the vexed topic of quantum cosmology that is fraught with its own problems of self-consistency. (Further discussion will be deferred until Chapter 16, where I shall suggest a quantum scenario which may solve the problem of the origin of the universe.) Accepting for now that, quantum theory notwithstanding, the total universe can be said to have a cause, what is that cause? God?

At this point we proceed to examine the second step of the cosmological argument: there cannot be an infinite chain of causes. The buck must stop somewhere. The galaxies form from swirling nebulae, the nebulae form from primeval hydrogen gas, the hydrogen forms from the protons created in the first brief bang, the protons were created out of spacewarps. The assumption has always been that this sequence must have a first member. Aquinas wrote:

> In the observable world causes are found to be ordered in series; we never observe, nor ever could, something causing itself, for this would mean it preceded itself, and this is not possible. Such a series of causes must however stop somewhere; for in it an earlier member causes an intermediate and the intermediate a last (whether the intermediate be one or many). Now if you eliminate a cause you also eliminate its effects, so that you cannot have a last cause, nor an intermediate one, unless you have a first. Given therefore no stop in the series of causes, and hence no first cause, there would be no intermediate causes either, and no last effect, and this would be an open mistake. One is therefore forced to suppose some first cause, to which everyone gives the name 'God'.[2]

In arguing against an infinite chain of cause and effect neither Aquinas nor Clarke object on the grounds that the chain is infinite as such. Indeed, both these thinkers developed their arguments in the context of an eternal, infinitely old universe, content to let the evidence for a creation rest on 'divine revelation' rather than rational argument. Rather, the objection seems to be that an infinite chain of cause and effect which encompasses the entire universe is allegedly impossible:

> If we consider such an infinite progression . . . 'tis plain this whole series of beings can have no cause from without, of its existence; because in it are supposed to be included all things that are or ever were in the universe; And 'tis plain it can have no reason within itself, of its existence; because no one being in this infinite succession is supposed to be self-existent or necessary . . . but every one dependent on the foregoing . . . An infinite succession therefore of merely dependent beings, without any original independent cause; is a series of beings that has neither necessity, nor cause . . . either within itself or from without: That is, 'tis an express contradiction and impossibility.[3]

The belief that an infinite succession of 'dependent beings' — loosely, an infinite chain of cause and effect — needs an explanation for its existence (which cannot be found when that chain includes all of existing things) has been sharply attacked by philosophers, especially Hume and Russell. In a famous B.B.C. debate with Father Copleston, Russell illustrated his point as follows: 'Every man who exists has a mother . . . but obviously the human race hasn't a mother.' In short, so long as each individual member of the succession is explained then, *ipso facto*, the succession is explained. And as every member of the chain owes its existence to some preceding member or members, each member of the infinite chain *is* explained. Asking for a cause of the whole universe has a different logical status from asking for a cause of an individual object or event within the universe.

In fact, the topic of 'sets of sets' is notoriously slippery. If a set is defined innocuously as any collection of things (concrete or abstract) then, as Russell showed by his famous paradox, a set of sets may not even be a set! Thus, we can envisage as a set a catalogue of all the books in a library. But is the catalogue itself to be included in this list? Sometimes it is. Call such catalogues 'Type I', and the others, that do not include themselves, 'Type II'. Now envisage as a set of sets a master catalogue at the central library. Its function is to list all Type II catalogues; it is a set of catalogues. Reasonable enough? Unfortunately not. The set of all Type II catalogues is paradoxical, as we discover as soon as we ask the question, is the master catalogue itself Type I or Type II? If Type II, then it does not include itself. But the master catalogue is defined as listing all self-excluding (Type II) catalogues. So it does list itself; it is Type I. But this cannot be so, for the master only lists Type II catalogues, so it cannot list itself if it is Type I. So it doesn't list itself; it is Type II. Result: self-contradictory nonsense.

The upshot of all this is that the concept of the entire universe of existing things is a subtle one indeed. It is not clear that the universe is a *thing*, and if it is defined as a set of things it runs the risk of paradox. Such issues lie in wait to ensnare all those who attempt to argue logically for the existence of God as a cause of all things.

Even granted the cosmological argument so far — that the universe must have a cause — there is a logical difficulty in attributing that cause to God, for it could then be asked 'What caused God?' The response is usually 'God does not need a cause. He is a *necessary* being, whose cause is to be found within himself.' But the cosmological argument is founded on the assumption that everything requires a cause, yet ends in the conclusion that at least one thing (God) does not require a cause. The argument seems to be self-contradictory. Moreover, if one is prepared to concede that something — God — can exist without an external cause, why go that far along the chain? Why can't the universe exist without an external cause? Does it require any greater suspension of disbelief to suppose that the universe causes itself than to suppose that God causes himself?

> If we stop, and go no farther (than God), why go so far? Why not stop at the material world? . . . By supposing it to contain the principle of its order within itself, we really assert it to be God.[4]

This quote of Hume is reminiscent of the vague belief of many scientists that 'God is nature' or 'God is the universe'.

Perhaps the most serious objection, however, to the causal version of the cosmological argument is the fact that cause and effect are concepts that are firmly embedded in the notion of time. Yet, as we have seen, modern cosmology suggests that the appearance of the universe involved the appearance of time itself. It is usually accepted that cause always precedes effect in time: the target shatters after the gun is fired, for example. In that case it is clearly meaningless to talk about God creating the universe in the usual causal sense, if that act of creation involves the creation of time itself. If there was no 'before' there can be no cause (in the usual sense) of the big bang, either natural or supernatural.

This point seems to have been well appreciated by St. Augustine (354–430) who ridiculed the idea of God waiting for an infinite time and then deciding at some propitious moment to create a universe. 'The world and time had both one beginning,' he wrote. 'The world was made, not in time, but simultaneously with time.'[5] This is a remarkable anticipation of modern scientific cosmology considering the completely erroneous ideas of space and time that were current in Augustine's day.

Curiously though, this profound interpretation of Genesis was later challenged when the Church came under the influence of the Ancient Greek tradition in the thirteenth century. In the ensuing controversy, the Fourth Lateran Council (1215), refuting Aristotle's philosophy of a universe of infinite age, insisted that, as an article of Christian faith, the universe *did* have a beginning in time, but even today theologians are still divided over the interpretation of the book of Genesis.

The problem about postulating a God who transcends time is that, though it may bring him into the 'here and now', many of the qualities which most people attribute to God only make sense within the context of time. Surely God can plan, answer prayers, express pleasure or anxiety about the course of human progress, and sit in judgement afterwards? Is he not continually active in the world, doing work, 'oiling the cogs of the cosmic machine' and so on? All of these activities are meaningless except in a temporal context. How can God *plan* and *act* except *in time*? Why, if God transcends time and so knows the future, is he concerned about human progress or the fight against evil? The outcome is already perceived by God. (We shall return to this topic in chapter 9.)

In fact, the very idea of God creating the universe is, as we have already seen, an act that takes place *in* time. When giving lectures on cosmology, I am often asked what happened before the big bang. The answer, that there was no 'before', because the big bang represented the appearance of time itself, is regarded with suspicion — 'Something must have caused it'. But cause and effect are temporal concepts, and cannot be applied to a state in which time does not exist; the question is meaningless.

If time really did have a beginning, any attempt to explain it in terms of causes must appeal to a wider conception of cause than that familiar to us in daily life. One possibility is to relax the requirement that cause always precedes effect. Is it possible for causes to act backwards in time, to produce prior effects? Of course, the idea of changing the past is replete with paradox. Suppose you could influence nineteenth-century events in such a way as to prevent your own birth, for example? Nevertheless there are a number of theories in modern physics that involve retro-active causation. Hypothetical faster-than-light particles (called tachyons) could accomplish this. To avoid paradox, one might suppose that the link between cause and effect is very loose and uncontrollable, or else it is of a more subtle variety. As we shall see, the quantum theory requires a sort of reversed time causality, inasmuch as an observation performed today can contribute to the construction of reality in the remote past. This point has been emphasized by the physicist John Wheeler: 'The quantum principle shows that there is a sense in which what the observer will do in the future defines what happens in the past — even in a past so remote that life did not then exist.'[6]

Wheeler here introduces mind ('the observer') in a fundamental way, as indeed one is obliged to do in the quantum theory, and involves the existence of mind at a later stage of cosmic evolution with the very creation of the universe:

> Is the very mechanism for the universe to come into being meaningless or unworkable or both unless the universe is guaranteed to produce life, consciousness and observership somewhere and for some little time in its history-to-be?[7]

Wheeler hopes that we can discover, within the context of physics, a principle that will enable the universe to come into existence 'of its own accord'. In his search for such a theory, he remarks: 'No guiding principle would seem more powerful than the requirement that it should provide the universe with a way to come into being.'[8] Wheeler has likened this 'self-causing' universe to a self-excited circuit in electronics.

Now even if it were possible to find a cause of the creation of spacetime from some later natural activity (be it mind or matter), it is hard to see how creation out of nothing could occur naturally. There would still have to be the 'raw materials' for mind or whatever to go to work on, retroactively. Wheeler suggests that space and time are indeed synthetic structures — they are made out of component 'bits' which he calls pregeometry. Many other physicists have suggested that space and time are not fundamental concepts, but approximations. Just as apparently continuous matter is in fact built out of atoms, so might spacetime be built out of more primitive, more abstract, entities. This might be one outcome of the attempt to find a quantum theory of gravity (gravity being merely spacetime geometry). Under extreme physical conditions, such as at the beginning of the big bang, spacetime might 'come apart' and the internal components be exposed. Expressing this in forward-time language, the big bang could have been the event when the 'cogwheels' engaged coherently and organized themselves into an apparently continuous spacetime. According to this view, the big bang was the beginning of space, time and matter, but not the limits of physics. Beyond the big bang (not 'before' for there was no before) lay the disorganized 'cogwheels' — physical things, but not *in* space or time.

Before leaving the topic of the creation and whether or not it is meaningful to ask if it was caused by something, we must consider the possibility that the answer may be yes, but that the something may not be God. As already remarked, the second part of the cosmological argument seeks to establish that a cosmic creator must indeed be God, but the discoveries of modern physics have opened up new possibilities of which the proponents of the cosmological argument could never have dreamed.

In the previous chapter it was explained how the creation of matter is adequately defined in terms of expanding space (spacewarp). Moreover, there seems to be no limit to the elasticity of space. The tiniest region can be expanded ad infinitum. At one billionth of a second after the creation the currently observed universe (all billion billion billion cubic light years of it) was shrunk to a volume about the size of the solar system. At earlier moments it was smaller than that. Hence space may grow out of nothing, and matter may come out of space. Nevertheless something, one feels, must start an infinitesimal blob of space on the path of explosive expansion, and this is where we get back to singularities, causation, and so on.

There is, however, an alternative explanation for our universe of space and matter. This can be dubbed, crudely speaking, the 'reproducing universe'. It is best described by analogy. As space is elastic, imagine it to be represented by a rubber sheet. (The sheet is only two-dimensional whereas space is three-dimensional. This is a conceptual shortcoming, but not a logical one. What is about to be described will also work in three dimensions, but is impossible to visualize in that case.)

Figure 5 shows a sequence of steps. First a bump is made in the rubber. Then the bump is inflated, all the while keeping the 'neck' very narrow where it connects to the sheet. The bump takes on the features of a balloon. Now allow the neck to shrink until the rubber touches and closes off the balloon completely. Finally cut the neck, releasing the balloon and allowing the neck to heal into a continuous sheet once more. The sheet has effectively given birth to a totally disconnected, independent sheet (balloon), which may then be inflated ad infinitum. If desired this new balloon could itself be used to generate other balloons.

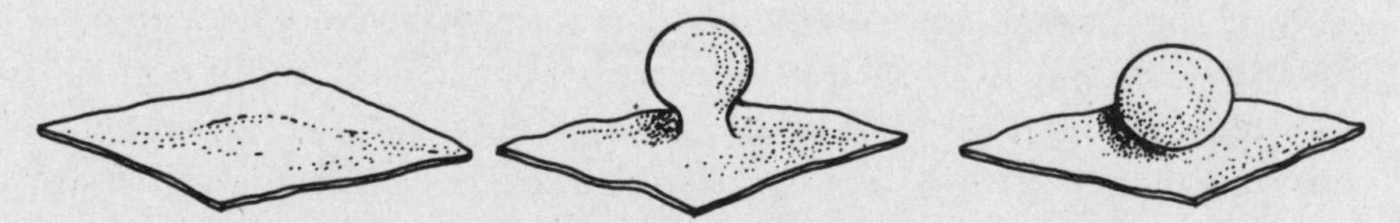

5 The elasticity of space suggested by Einstein's general theory of relativity permits the growth and separation of a 'daughter universe' (bubble) from the 'parent universe' (sheet). Such topology changes have been proposed in some recent theories, but are not at all well understood.

If we envisage our universe — all of the space to which we can possibly have physical access — as the 'new balloon' then it is certainly the case that this universe has not always existed: it was created. However, its creator can still be found within the scope of natural physical processes, namely a creation mechanism with its origin in the 'mother sheet'. That sheet is now totally inaccessible to us, it is beyond our spacetime, so we can find no cause within our universe for its existence, and yet God is not involved.

The central feature which emerges from this idea is that what is usually regarded as 'the universe' might in fact be only a disconnected fragment of spacetime. There could be many, even an infinite number of other universes, but all physically inaccessible to the others. With this definition of 'universe' the explanation for our cosmos does not lie within itself — it lies beyond. It does not involve God, only spacetime and some rather exotic physical mechanism.

Such a mechanism has been proposed recently in a number of theoretical studies.[9] Under extreme heat it is conceivable that space could become unstable to 'breeding' other 'balloons' in this way. One could even envisage a sufficiently advanced technological community deliberating engineering the creation of new universes. Nevertheless, purists will no doubt object that this hypothesis of the creation constitutes only a pseudo-explanation, for it still does not account for the totality of 'sheets and bubbles'. That is true, but the example does serve to illustrate that everything that we can, in principle, perceive in our universe may still have been created by *natural* causes a finite time ago, and that what (if anything) lies outside all our spacetime may not be entirely supernatural.

What, then, has this analysis contributed to our search for God the creator? The argument that there must exist a first cause of everything is open to serious doubt so long as we adhere to any simple notion of cause, irrespective of whether the universe is infinitely old, or had a definite beginning in time. Exotic causal mechanisms, such as reversed-time causality or quantum mental processes might conceivably remove the need for a prior cause of the creation. Nevertheless one is still left with a feeling of unease. The theologian Richard Swinburne writes:

> It would be an error to suppose that if the universe is infinitely old, and each state of the universe at each instant of time has a complete explanation in terms of a previous state of the universe and natural laws (and so God is not invoked), that the existence of the universe throughout infinite time has a complete explanation, or even a full explanation. It has not. It has neither. It is totally inexplicable.[10]

To illustrate the point, suppose that horses had always existed. The existence of each horse would be causally explained by the existence of its parents. But we have not explained yet why there are horses at all — why there are horses rather than no horses, or rather than unicorns, for example. Although we may be able to find a cause for every event (unlikely in view of quantum effects), still we would be left with the mystery of why the universe has the nature it does, or why there is any universe at all.

4. Why is there a universe?

> 'There is a reason in Nature why something should exist rather than not.'
>
> Leibniz

> 'The more the universe seems comprehensible, the more it also seems pointless.'
>
> Steven Weinberg

The idea of God-the-creator, who caused the universe to come into being of his own free will, is firmly rooted in the Christian-Judaic culture. Yet we have seen how such an assumption raises more problems than it solves, and has been seriously questioned by theologians for centuries. The difficulty involves the nature of time. Today we know that time is linked inseparably to space, and that space-time is as much a part of the physical universe as matter. As we shall see in Chapter 9, time has its own laws of change and behaviour; it is demonstrably part of physics.

If time belongs to the physical universe, and is subject to laws of physics, it must be included in the universe that God is supposed to have created. But what does it mean to say that God *caused* time to come into existence, when by our usual understanding of causation a cause must precede its effect? Causation is a temporal activity. Time must already exist before anything can be caused. The naive image of God existing *before* the universe is clearly absurd if time did not exist — if there was no 'before'.

These difficulties were already apparent, as we have seen, to St. Augustine in the fifth century. They were articulated especially by

Boethius a century later, and developed into a concept of 'creation' that is far more abstract and subtle than the one which is still familiar to most laymen. According to this refined viewpoint, God exists entirely outside space and time; in some sense 'above' nature, rather than before it. The concept of a timeless God is not an easy one, and I shall defer the main discussion of this topic until Chapter 9 which deals with the nature of time in more depth.

The God who is outside time is regarded as 'creating' the universe in the more powerful sense of 'holding it in being at every instant'. Instead of God simply starting the universe off (a belief known as deism rather than theism), a timeless God acts at all moments. The remote cosmic creator is thus given a greater sense of immediacy — he is acting here and now — but at the expense of some obscurity, for the idea of God being above time is a subtle one.

The alternative roles of God in time, causing the creation, and a timeless God holding the universe (including time) in being, are sometimes illustrated schematically in the following way.[1] Imagine a sequence of events, each one causally dependent on the preceding one. They can be denoted as a series . . . E_3, E_2, E_1, stretching back in time. Thus, E_1 is caused by E_2, which in turn is caused by E_3 and so on. This causal chain can be denoted as follows:

$$\ldots \rightarrow E_4 \xrightarrow{L} E_3 \xrightarrow{L} E_2 \xrightarrow{L} E_1$$

where the 'L's remind us that one event causes the next through the operation of the laws of physics, L.

The concept of a causal God (which we considered in detail in the previous chapter) can then be illustrated by making God, denoted G, the first member of this series of causes:

$$G \rightarrow \ldots \rightarrow E_4 \xrightarrow{L} E_3 \xrightarrow{L} E_2 \xrightarrow{L} E_1$$

By contrast, if God is outside time, then he cannot belong to this causal chain at all. Instead, he is above the chain, sustaining it at every link:

$$\ldots \rightarrow E_4 \xrightarrow{\overset{G}{\downarrow}\,L} E_3 \xrightarrow{\overset{G}{\downarrow}\,L} E_2 \xrightarrow{\overset{G}{\downarrow}\,L} E_1$$

and this picture could apply equally well whether the chain of causes has a first member (i.e. a beginning in time) or not (as in an infinitely old universe). With this picture in mind, we may say that God is not so much a cause of the universe as an *explanation*.

These ideas are not easy to grasp. Roughly speaking, the laws of physics are apparent to us as regularities in the way things happen: the precision motion of the planets in their orbits, the orderly pattern of lines in the spectrum of an element, and so on. When we press the brake pedal in a moving car we expect the car to slow down. When we ignite gunpowder we expect it to explode. We expect a hot flame to melt a block of ice, or a hard floor to smash a falling vase. The world is not haphazard and chaotic but, at least to a certain extent, predictable and orderly.

From our limited perspective within spacetime we interpret these regularities in terms of cause and effect: the sun's gravity causes the Earth's orbit to curve, and so on. But there is an alternative possibility — that every event is actually caused by God, operating on our universe from outside, carefully arranging the events to display the regularities.

There is a helpful analogy here. Imagine a machine-gunner facing a target screen. As he fires the gun, he sweeps his aim at a steady rate from side to side. The end result is a pattern of equispaced bullet holes. Now a two-dimensional creature obliged to live permanently in the flatland of the screen would perceive this sequence of events as the regular appearance of holes in his world. With careful observation he would deduce that the holes are not formed at random, but periodically, and moreover they are arranged in a geometrically simple way, with equal distance between them. Confidently this flatlander would proclaim a new law of flatland physics: the law of hole creation. He would conclude that the appearance of each hole *causes* the appearance of the next in line, in a regular way. After all, one hole is always followed by another in a simple sequence. From the limited perspective of his two-dimensional world, the flatlander misses entirely the fact that the holes are actually *completely independent* of each other, and the regularity in their arrangement is due entirely to the activity of the machine-gunner. In the same way, the orderly operation of the cosmos can be explained by God creating each event in spacetime in an organized way from some wider perspective. A higher dimensional space? A physical structure which is not space? An entirely non-physical structure (whatever that may mean)?

What is the justification for this belief? Look around you. See the complex structure and elaborate organization of the universe. Puzzle over the mathematical formulations of the laws of physics. Stand perplexed before the arrangement of matter, from the whirling galaxies to the beehive activity of the atom. Ask why these things are the way they are. Why *this* universe, *this* set of laws, *this* arrangement of matter and energy? Indeed, why anything at all?

Every thing and every event in the physical universe must depend for its explanation on something outside itself. When a phenomenon is explained, it is explained in terms of *something else*. But if that phenomenon is all of existence — the entire physical universe — then clearly there is nothing *physical* outside the universe (by definition) to explain it. So any explanation must be in terms of something non-physical and supernatural. That something is God. The universe is the way it is because God has *chosen* it to be that way. Science, which by definition deals only with the physical universe, might successfully explain one thing in terms of another, and that in terms of another and so on, but the totality of physical things demands an explanation from *without*.

This line of reasoning, which takes as its basis the assertion that all physical things are contingent upon something else, is known as the contingency argument and is the second version of the cosmological argument for the existence of God. It is open to some of the criticisms which have been deployed against the other version of the cosmological argument (the causal argument considered in the previous chapter).

In a sense that contingency argument falls a victim of its own success, for suppose we enlarge the definition of 'universe' to include God. What, then is the explanation for the total system of God plus the physical universe of space, time and matter? In short, what explains God? The theologian answers: 'God is a *necessary* being, without need of explanation; God contains within himself the explanation of his own existence.' But does this mean anything? And if it does, why can't we use the same argument to explain the universe: The universe is *necessary*, it contains within itself the reason for its own existence? Indeed, that seems to be Wheeler's position described in the last chapter.

The idea of a physical system containing an explanation of itself might seem paradoxical to the layman but it is an idea that has some precedence in physics. While one may concede (ignoring quantum effects) that every event is contingent, and depends for its explanation on some other event, it need not follow that this series either continues endlessly, or ends in God. It may be closed into a loop. For example, four events, or objects, or systems, E_1, E_2, E_3, E_4, may have the following dependence on each other:

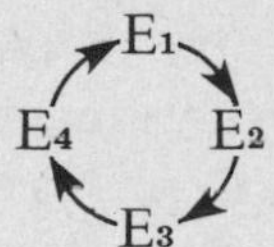

A theory of precisely this sort was once popular with some particle physicists in their attempts to explain the structure of matter. Here there is a well-known chain of explanation: matter is made of molecules, which are made of atoms, which are made of electrons and nuclei, which are made of protons and neutrons. There has been a widespread belief (since ancient Greece) that this chain of explanation will have an end; that there exists a small number of truly elementary particles that have no internal parts and which are the building blocks of all matter. If we can but probe into ever smaller regions within the atom, sooner or later these fundamental, structureless particles will be discovered. At present, this theory receives strong experimental support in the shape of the so-called quark theory (see Chapter 11).

An alternative picture, permitted by the weird properties of the quantum theory, is that (in a subtle sense to be clarified in later chapters) there are no elementary particles at all. Instead, every particle (at least every subnuclear particle) is made up of every other. No particle is elementary or primitive, but each contains something of the identity of all the others. The idea of a system of particles generating themselves in a self-consistent loop of explanation is reminiscent of the story of the boy who fell into a bog and hauled himself out by pulling on his own bootstraps, so physicists call such modes of explanation 'bootstrapping'. One could envisage a 'bootstrap universe' containing its own explanation entirely in terms of natural, physical interactions.

But surely, counters the theologian, God, who is infinitely powerful and infinitely knowledgable and hence the *simplest* being that one can evisage, is more likely to contain the reason for his own existence than is the universe, which is *complicated* and *special* in its many particular features:

> There is quite a chance that if there is a God he will make something of the finitude and complexity of a universe. It is very unlikely that a universe would exist uncaused, but rather more likely that God would exist uncaused. The existence of the universe is strange and puzzling. It can be made comprehensible if we suppose that it is brought about by God. This supposition postulates a simpler beginning of explanation than does the supposition of the existence of an uncaused universe, and that is grounds for believing the former supposition to be true.[2]

This counter is very persuasive. It takes a lot to believe that this intricate universe with so many characteristic, contingent features, just happens to be. Can we really accept it as a brute, inexplicable fact? Yet a single, simple, infinite mind (though the logic of even its existence may be perplexing to us) seems an altogether more plausible candidate for something that exists of necessity.

The scientist, however, may wish to challenge the assumption that an infinite mind (God) is simpler than the universe. In our experience, mind only exists in physical systems that are above a certain threshold of complexity. The brain is a highly complicated system. (In Chapter 6 we shall see that mind must be regarded as a 'holistic' concept — a pattern of activity.) While it is possible to imagine a disembodied mind, there must be some means of expression of the pattern, and the pattern itself is complex. So it could be argued that an infinite mind is infinitely complex and hence far *less* likely than a universe, many parts of which are far too short on complexity to support a mind.

Perhaps, then, God is not a mind, but something simpler? Does it in any case make sense to talk about a mind existing timelessly? Aren't thoughts, decisions, and so on things that take place in time? But if God cannot *decide* (or hope, or judge, or converse) in what sense is he responsible for the nature and existence of the universe? Is such a being anything that we would recognize as a God at all? Despite these doubts, we are still left with the complexity and specificity of the universe to account for. Why *this* universe?

This is the question I shall take up more fully in Chapter 12, but here we may note what I believe to be the central issue in assessing the relative plausibility of a self-caused universe as against requiring God for its explanation. In the foregoing discussion it was taken for granted that the universe is very complicated, and that God provides a ready explanation for its features. But has the universe always been complicated? Could this complexity not have arisen naturally as a result of perfectly ordinary physical laws?

According to our best scientific understanding of the primeval universe it does indeed seem as though the universe began in the simplest state of all — thermodynamic equilibrium — and that the currently-observed complex structures and elaborate activity only appeared subsequently. It might then be argued that the primeval universe is, in fact, the simplest thing that we can imagine. Moreover, if the prediction of an initial singularity is taken at face value, the universe began in a state of infinite temperature, infinite density and infinite energy. Is this not at least as plausible as an infinite mind?

The success of the above argument depends crucially on whether it can be demonstrated that cosmic complexity and order really have arisen spontaneously from the simple primeval state. At first sight this claim seems to be in flagrant contradiction with the second law of

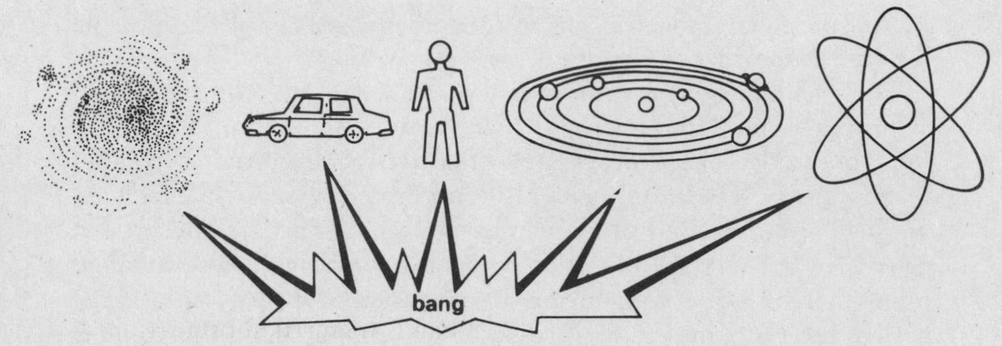

6 Mystery surrounds how order has emerged out of chaos in the universe. The present orderly structures and complex activity has somehow arisen from the featureless ferment of the big bang, in apparent defiance of the second law of thermodynamics which requires that order decreases, rather than increases, with time. The resolution of the paradox may concern the peculiar properties of gravity

thermodynamics, which requires just the opposite — that order gives way to disorder, so that complex structures tend to decay to a final state of disorganized simplicity. Thus, E.W. Barnes wrote in the 1930s:

> In the beginning there must have been a maximum organization of energy . . . In fact, there was a time when God wound up the clock (i.e. the cosmic mechanism) and a time will come when it will stop if He does not wind it up again.[3]

We now know that this is wrong. The primeval state was not one of maximum organization but one of simplicity and equilibrium. The apparent conflict of this fact with the second law has only recently been resolved.

The problem is that the second law strictly applies only to isolated systems. Now it is physically impossible to isolate anything from gravity — there are no gravity shields, and even if there were the system concerned could not escape its own gravity. In the expanding universe, the cosmic material comes under the influence of the cosmological gravitational field — the cumulative gravity of the rest of the universe. This coupling to gravity opens the way to the injection of order into the cosmic material by the gravitational field. We know that, given an external supply of energy, order can be created in one system at the expense of disorder in another. Thus the flux of heat and light from the sun generates the highly complex order of the Earth's biosphere, but only by sacrificing irreversibly the limited fuel resources of the solar core. In the same way, an expanding universe can generate order in the cosmic material.

A very simple example can be given of how the expansion of the universe can be used in place of God to 'wind up the clock'. It has already been remarked that the primeval cosmic substance was very hot, but the expansion of the universe caused it to cool. An elementary scaling argument yields the temperature of the substance at each stage of the expansion. However, the temperature will depend to some extent on the nature of the substance itself. In the case of radiant heat (electromagnetic energy), the temperature declines in proportion to the size of a typical expanding region of space: double the size and the temperature is halved. On the other hand, material substance, such as hydrogen gas, cools much faster, like the square of the size. This implies that, so long as hydrogen gas is decoupled from radiant heat, the expanding universe will cause a temperature difference to open up between these two components of the cosmic substance. As any engineer knows, a temperature differential is an ideal source of useful energy, and is in essence the secret of the sun's power to generate life on Earth. Thus, the expansion of the universe is capable of creating order where none existed before.

Using analyses like this, it is possible to trace, step by step, the origin of most of the orderly structure that we observe in the universe today back to the expansion of the universe in the primeval era.[4] The above cited example is actually not the most important. By far the greatest source of organized energy today is the highly reactive hydrogen gas which constitutes about seventy-five per cent of the cosmic material. Hydrogen provides the fuel for all normal stars. When it is burned (in nuclear fusion reactions) it is ultimately converted to heavier elements such as iron. Iron is just nuclear ash; it has no useful nuclear energy locked up inside. We therefore owe the existence of the stellar order to the preponderance of hydrogen over iron.

This circumstance can be explained by the cosmic expansion. In the primeval phase it was too hot for any composite nuclei (such as iron) to exist. Only hydrogen nuclei (individual protons) — the simplest substance — could survive. With the continual expansion and cooling, the way lay open for the conversion of hydrogen into heavier elements and, as discussed in the previous chapter, the cosmic material made some progress down this road. It did not, however, get very far. About twenty-five per cent reached helium (the next simplest element) and only a minute fraction beyond. The blame for the aborted journey can be laid at the door of the expansion. It was far too rapid to give the material enough time to undergo all the complex nuclear reactions necessary before heavy, composite, nuclei like iron can be synthesized. After only a few minutes of 'cooking' the temperature had sunk below the threshold needed for nuclear reactions to ignite. The nuclear fire went out, leaving most of the material 'frozen' in the form of hydrogen or helium. Only with the formation of stars, which occurred much later on, were local hot-spots created in which the journey could be resumed.

In conclusion, it appears that in an expanding universe organized energy can appear spontaneously, without the necessity for it being present at the outset. There is then no need to attribute the cosmic order (low entropy) either to the activity of a Deity or to the input of organization at the initial singularity. The singularity could have coughed out totally random and chaotic energy, which then organized itself spontaneously into the present arrangement under the influence of the expanding universe. Notice that now we not only have attributed the origin of matter to expanding space (see page 32), but also the origin of its organization.

This cannot, however, be the whole story. The gravitational field, which is ultimately responsible for generating order via the cosmic expansion, presumably suffers some disordering tendency as a result. Thus we can explain the order of material things by shifting the responsibility on to gravity, but then we have to explain how the order appeared in the gravitational field in the first place. Where does the buck stop?

The issue turns on whether or not the second law of thermodynamics applies to gravity as well as to matter. Nobody really understands this. Recent work on black holes suggests that it does, but different physicists have drawn opposite conclusions (see Chapter 13). Some, such as Roger Penrose, conclude that the large scale cosmic gravitational field is in a very low entropy (highly ordered) state which therefore requires an input of order at the creation. Others, such as Stephen Hawking, claim that the cosmic gravity is highly disordered, and is the expected result of purely random and unstructured influences emerging from the initial singularity. Because no one yet knows how to quantify the orderliness of a spacewarp (i.e. gravity) the issue remains undecided. Nevertheless, the debate illustrates an important point. Future progress in theoretical physics might well clarify the concepts involved, and enable a definitive statement to be made as to whether the universe was created with or without order. Thus may science come one day to answer a question that has long occupied the attention of theologians and philosophers.

Whatever the outcome of the debate about quantifying the entropy of gravity, one curious thing has already emerged. In systems such as boxes of gas, where gravity is so small it can be ignored, low entropy (ordered) states are complicated, while high entropy (disordered) states are simple. For example, a box in which all the gas molecules are crowded into the corners clearly involves a more complicated arrangement than the equilibrium (maximum entropy) state in which the gas is distributed uniformly throughout the box. By contrast, a low entropy gravitating system is geometrically much *simpler* than one in a high entropy state. Gravity has a tendency to grow structures spon-

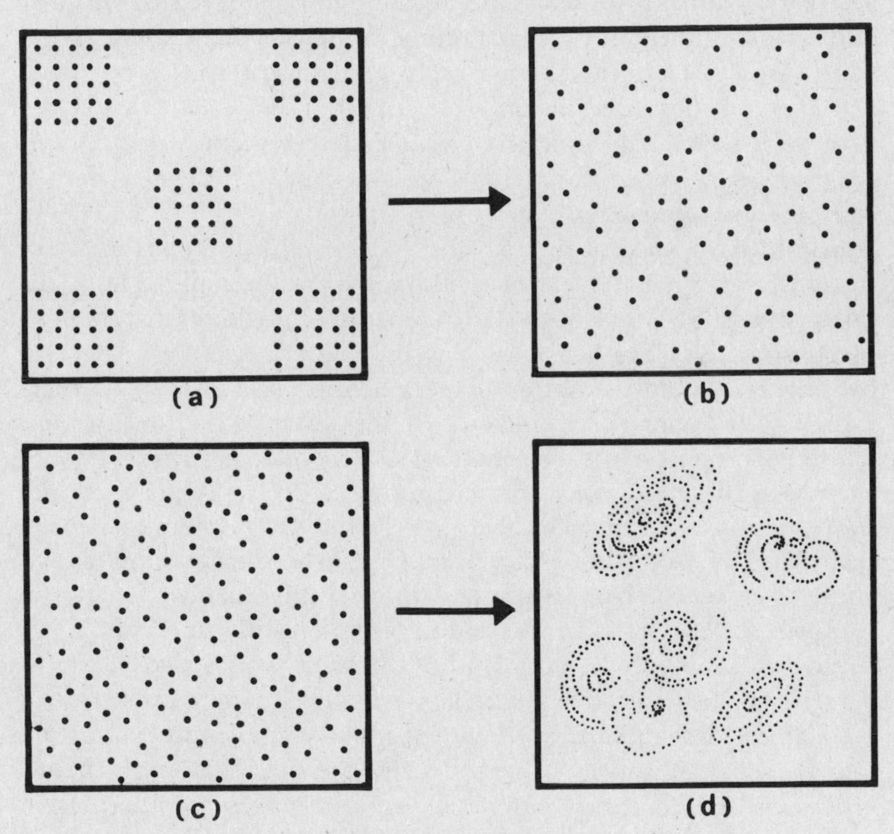

7 The concept of order depends crucially on whether gravity can be ignored. Box (a) contains a gas for which gravity is negligible. Its highly ordered molecular arrangement soon gives way to featureless disorder (maximum entropy) as a result of molecular agitation and collisions. The final state is shown in (b). By contrast, a gravitating 'gas', for example a system of stars, will do just the opposite. The initial uniform configuration (c) will tend to fragment and become clumpy as the stars fall together and organize themselves into clusters (c.f. galaxies). The ultimate result of this clumping would be a number of black holes.

taneously. Thus, a uniform distribution of matter (stars, or gas) will tend to grow clumpy with time, forming into clusters and dense accumulations. In summary, for non-gravitating systems order means complexity and disorder means simplicity. For gravity it is the other way round (see Fig. 7).

If the universe really did start out with a highly ordered, low entropy, gravitational field, then this field would have been smooth and uniform. So we see that it is possible, in the special case of gravity, to satisfy both the requirement of simplicity, and the requirement of low initial entropy (order). This means we can regard the *simplest* universe (a uniform one) as containing immense potential for generating complexity later. This is surely a pleasing result. If we were expected to believe that the universe appeared uncaused, what better than for it to have the simplest possible arrangement of matter and gravity, yet without compromising its ability to develop into a complex and interesting form subsequently?

In spite of this success, there is more to the world than just the *state* of the universe. What about the *laws*? Granted that, initially at least, the universe was in a very simple state, there can be no doubt that the laws of physics are still rather numerous and special. Are these laws not contingent? Could we not envisage a host of alternatives? Furthermore, what about the *constituents* of the universe — the protons, neutrons, mesons, electrons, and so on. Why *those* particles? Why do they have the masses and charges that they do? Why are there not more, or less, types of such subatomic particles? The theologian has a ready answer. God made it that way. God, who is infinite simplicity, chose to create the laws of physics and the constituents of matter in complex variety, in order to produce an interesting universe.

Now it is only very recently that the scientist has also begun to perceive an answer to this point. The new developments arise from a programme of theoretical work aimed at unifying the forces of nature into a single descriptive scheme. According to this theoretical scheme, which will be described more fully in a later chapter, the present profusion of physical laws is purely a low-temperature phenomenon. As the temperature of matter is raised, so the varied forces that act upon it begin to merge their identity until, at the staggering temperature of 10^{32}K (that is a hundred thousand billion billion billion degrees absolute) all the forces of nature should merge into a single superforce with a remarkably simple mathematical form. Furthermore, all the many disparate subatomic particles lose their identities too, their varied characteristics disappearing in the searing heat. Evidence for this convergence to simplicity comes from years of study of high-energy physics (high energy is the same as high temperature in this context). Physicists tend to find that as the energy is raised, so complex subatomic structures break apart to reveal simpler constituents, and complicated forces become simpler in operation.

If these ideas are right — and it is premature to conclude more than that the signs are encouraging — then they have profound implications for the big bang theory. In the unlimited temperatures of the creation, only the superforce would have operated, with a handful of simple particle species. The current differentiated forces and particles would only have emerged as the universe cooled. Thus, the state of the universe, the laws of physics and the constituents of matter all seem to have started out in an exceedingly simple form.

Still, the sceptical theologian will reply, even a single superforce and a handful of simple particles require an explanation. Why that particular superforce? In fact, why any *law* at all?

This is a point that we shall return to in the final chapter. Some physicists, inspired by the simplicity of nature's fundamental laws, have argued that perhaps the ultimate law (in this case the superforce) has a mathematical structure which is uniquely defined as the only logically consistent physical principle. That is to say, physics is proclaimed 'necessary' in the same way that God is proclaimed necessary by theologians. Should we then conclude that *God is physics* as some philosophers (such as Plato) seem to have done?

A few physicists, notably Stephen Hawking, have argued that a remarkably simple primeval state of the universe is, in fact, to be expected.[5] The reason for this concerns the initial singularity, discussed briefly in Chapter 2. The essential feature of a singularity is that it is rather like an edge or boundary to spacetime and hence, one supposes, to the physical universe. An example of a singularity is the infinitely dense, infinitely compact state that marked the beginning of the big bang. Singularities are also expected to occur inside black holes and perhaps elsewhere as well.

Because all our physical theories so far are formulated in the context of space and time, the existence of a boundary to spacetime suggests that natural physical processes cannot be continued beyond such a thing. In a fundamental sense a singularity represents, according to this view, the outer limits of the natural universe. At a singularity, matter may enter or leave the physical world, and influences may emanate therefrom that are totally beyond the power of physical science to predict, even in principle. A singularity is the nearest thing that science

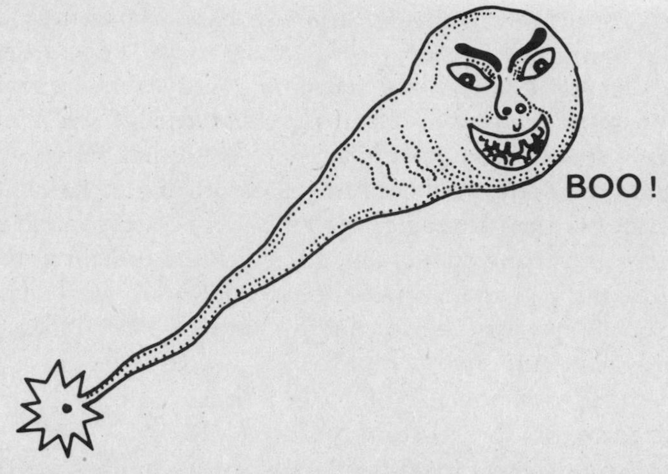

8 A singularity (dot) represents the ultimate unknowable in science. It is an edge or boundary of spacetime at which matter and influences can enter or leave the physical universe in a totally unpredictable fashion. If a singularity is 'naked' then anything can apparently come out of it without prior physical causation. Some cosmologists believe that the universe emerged without cause from a type of naked singularity. If these ideas are correct, a singularity is the interface between the natural and the supernatural.

has found to a supernatural agent.

For many years it was thought that singularities were an artefact due to over-idealization in the gravitational model used. Then, in a series of brilliant and embracing mathematical theorems, Penrose and Hawking proved that singularities were quite general and, under all reasonable physical conditions, unavoidable, once gravity becomes strong enough. It was certainly strong enough in the big bang.

Since they had to be taken seriously a lot of thought was expended on how singularities might behave. The choices boil down to this. What comes out of a singularity is either totally chaotic and unstructured, or it is coherent and organized. In the former case, the big bang singularity simply coughs out a randomly arranged universe displaying no particular order, and in the latter case the universe emerges with a degree of organization present, wound up and ready for action.

Hawking has proposed a 'principle of ignorance' which says that the singularity is the ultimate unknowable, and therefore should be totally devoid of information (in physics, information is roughly the same as order — the negative of entropy).[6] Hence, anything that emerges from a singularity is totally random and chaotic. This accords well with the belief that the primeval universe was in a state of maximum disorder (thermodynamic equilibrium).

Many of these ideas are at the frontier of modern theoretical physics and will only be clarified by future developments. There is no unanimous agreement among physicists about the status of spacetime singularities, or even about the precise state of the primeval universe. Yet the flow of ideas generated by recent advances in scientific cosmology has undoubtedly regenerated and given a new slant to the debate about God and the existence of the universe.

5. What is life? Holism vs reductionism

'So God created man in his own image.'

Genesis 1: 27

'We are survival machines — robot vehicles blindly programmed to preserve the selfish molecules known to us as genes.'

Richard Dawkins in *The Selfish Gene*

According to the theologian, life is the supreme miracle, and human life represents the crowning achievement of God's cosmic masterplan. To the scientist, life is the most astonishing phenomenon in nature. A hundred years ago, the subject of the origin and evolution of living systems became the battleground for history's greatest collision between science and religion. Darwin's theory of evolution shook the foundations of Christian doctrine and, more than any other pronouncement since Copernicus placed the sun at the centre of the solar system, brought home to ordinary people the far-reaching consequences of scientific analysis. Science, it appeared, could change man's entire perspective of himself and his relation to the universe.

This is primarily a book about physics, and we shall not dwell in detail on the Darwinian revolution, its repercussions for the Church, or the curious resurrection of anti-Darwin sentiment in the recent 'creationist' movement. All those topics have been thoroughly documented elsewhere. Instead, this chapter will examine the physicist's view of living organisms, and address the questions: what is life and does it provide evidence for a divine spirit?

The Bible states quite explicitly that life is the direct result of God's activity; it did not arise naturally as a result of ordinary physical processes established after the creation of heaven and Earth. Instead, God chose to produce, by divine power, first the plants and animals, then Adam and Eve. Of course, the vast majority of Christians and Jews now concede the allegorical nature of Genesis, and make no attempt to defend the biblical version of the origin of life as historical fact. Nevertheless, the divine nature of life — especially human life — continues to be a central feature of contemporary religious doctrine.

Is life divine? Did God literally manipulate molecules of non-living matter in violation of the laws of physics and chemistry to produce, miraculously, the first living thing? Did he further manipulate the genetic structure of some ape-like creature thousands (or millions) of years ago to produce man? Or is life the result of purely natural, if complex, physical and chemical activity, and man the end product of a long and convoluted evolutionary development? Can life be created artificially, in the laboratory, or must it contain an added ingredient — a divine spark — before it can be viable?

What is life? To the physicist the two distinguishing features of living systems are *complexity* and *organization*. Even a simple single-celled organism, primitive as it is, displays an intricacy and fidelity unmatched by any product of human ingenuity. Consider, for example, a lowly bacterium. Close inspection reveals a complex network of function and form. The bacterium may interact with its environment in a variety of ways, propelling itself, attacking enemies, moving towards or away from external stimuli, exchanging material in a controlled fashion. Its internal workings resemble a vast city in organization. Much of the control rests with the cell nucleus, wherein is also contained the genetic 'code', the chemical blueprint that enables the bacterium to replicate. The chemical structures that control and direct all this activity may involve molecules with as many as a million atoms strung together in a complicated yet highly specific way. Fundamental to the chemical basis of life are the nucleic acid molecules, RNA and DNA, with their famous 'double helix' architecture.

It is important to appreciate that a biological organism is made from perfectly ordinary atoms. Indeed, part of its metabolic function is to acquire new substances from the environment and to discard degenerated or unwanted substances. An atom of carbon, hydrogen, oxygen, or phosphorus inside a living cell is no different from a similar atom outside, and there is a steady stream of such atoms passing into and out of all biological organisms. Clearly, then, life cannot be reduced to a property of an organism's constituent parts. Life is not a cumulative phenomenon like, for example, weight. For though we may not doubt that a cat or a geranium is living, we would search in vain for any sign that an individual cat-atom or geranium-atom is living.

Sometimes this appears paradoxical. How can a collection of inanimate atoms be animate? Some people have argued that it is impossible to build life out of non-life, so there must be an additional, non-material, ingredient within all living things — a life-force — or spiritual essence which owes its origin, ultimately, to God. This is the ancient doctrine of vitalism.

An argument frequently used in support of vitalism concerns behaviour. A characteristic feature of living things is that they appear to behave in a purposive way, as though towards a specific end. This goal-oriented or 'teleological' quality is most evident in the higher life forms, but even a bacterium can give the impression of striving to achieve certain rudimentary tasks, such as the acquisition of food.

In the 1770s Luigi Galvani discovered that a frog's muscle twitched when touched with a pair of metal rods, and concluded that this 'animal electricity' was none other than the occult spirit of life at work. Indeed, the belief that electricity is somehow connected with the life-force lives on in the story of Frankenstein, a man-made monster brought to life amid the crackling discharge of an electrical contrivance.

In more recent years some investigators of the so-called paranormal claim to have directly detected the mysterious life-force by a somewhat improbable combination of psychic power and advanced technology. Indistinct photographs showing nebulous or filamentary rays and blotches emanating from a variety of living things (including people's fingers) have been exhibited.

Unfortunately, it is hard to find any real scientific support for these vague conjectures. Apparently the only way in which the hypothetical life-force manifests itself is through life; living things display the life-force, non-living things do not. But this reduces the life-force to a mere word, not an explanation for life. For what does it mean to say that a person, or a fish, or a tree, has the life-force? Only that it is living. As for the manifestation of the life-force in obscure and mysterious 'experiments', these episodes are notorious for their non-repeatability, and are so obviously open to the challenge of fraud that they are taken seriously by very few professional scientists.

The mistake made in invoking a life-force is to overlook the fact that a multi-component system may possess collectively qualities that are absent, or meaningless, for the individual components. To take another example, consider a newspaper picture of a face composed from myriads of tiny dots. No amount of scrutiny of the individual dots will reveal a face. Only by standing back and viewing the collection of dots as a whole, on a coarser scale, does the image emerge. The face is not a property of the dots *themselves*, but of the *collection* together — it is to be found in the pattern, not in the constituents. Likewise, the secret of life will not be found among the atoms themselves, but in the pattern of their association — the way they are put together, in the information encoded within the molecular structures. Once the existence of collective phenomena is appreciated, the need for a life-force is removed. Atoms do not need to be 'animated' to yield life, they simply have to be arranged in the appropriate complex way.

The distinction being made here is sometimes referred to as 'holism' versus 'reductionism'. The main thrust of Western scientific thinking over the last three centuries has been reductionist. Indeed the use of the word 'analysis' in the broadest context nicely illustrates the scientist's almost unquestioning habit of taking a problem apart to solve it. But of course some problems (such as jigsaws) are only solved by putting them together — they are synthetic or 'holistic' in nature. The picture on a jigsaw, like the speckled newspaper image of a face, can only be perceived at a higher level of structure than the individual pieces — the whole is greater than the sum of its parts.

Scientific reductionism began in earnest with the physics of the nineteenth century and the development of the atomic theory of matter. It is a path followed more recently by biologists who have achieved such notable successes in unravelling the molecular basis of life. These pace-setting advances have encouraged a reductionist approach from many other areas of human enquiry.

The evils of rampant reductionism have attracted sharp criticisms, however. The writer, Arthur Koestler, remarks: 'By denying a place for values, meaning and purpose in the interplay of blind forces, the reductionist attitude has cast its shadow beyond the confines of science, affecting our whole cultural and even political climate.'[1] Many critics complain that attempts to explain away living organisms as nothing but meaningless mounds of atoms, formed pointlessly as the result of random accidents, seriously devalues our own existence.

The British neurobiologist, Donald MacKay, who is a well-known defender of Christian doctrine, challenges what he calls this 'nothing-buttery' attitude, so prevalent among contemporary biologists. In *The*

Clockwork Image he cites, as an illustration of his argument, the operation of one of those familiar advertising displays consisting of hundreds of electric lamps that flash on and off in sequence to spell out a message. An electrical engineer could give a complete and accurate description of this system in terms of electric circuit theory, explaining exactly why and how each light is flashing. Yet the claim that the advertising display is therefore nothing but electrical pulses in a complex circuit is absurd. True, the electrical description is neither wrong nor incomplete at its own particular level of description, but it makes no mention of the message. The concept of the message is outside the terms of reference of the engineer's job. It only becomes apparent when the operation of the display *as a whole* is considered. We may say that the message is on a higher level of structure than circuits and lamps: it is a holistic feature.

In the case of living systems, nobody would deny that an organism is a collection of atoms. The mistake is to suppose that it is nothing but a collection of atoms. Such a claim is as ridiculous as asserting that a Beethoven symphony is nothing but a collection of notes or that a Dickens novel is nothing but a collection of words. The property of life, the theme of a tune or the plot of a novel are what have been called 'emergent' qualities. They only emerge at the collective level of structure, and are simply meaningless at the component level. The component description does not contradict the holistic description; the two points of view are complementary, each valid at their own level. (When we come to examine the quantum theory, we shall encounter again the idea of two different, complementary, descriptions of a single system.)

The importance of level distinction is very familiar to computer operators. A modern electronic computer consists of an intricate network of electrical circuits and switches around which passes a complex sequence of electrical pulses. This is the hardware level of description. On the other hand, the same electrical activity might represent the solution of a set of mathematical equations or the analysis of a missile trajectory. Such a description, which is at a higher level than hardware, uses concepts such as programs, operations, symbols, input, output, answer, among others, which are meaningless at the hardware level. A component switch inside the computer does not fire in order to compute, a square root, for example. It fires because the voltage is right and the laws of physics oblige it to. The high-level program description of computer operation is called the software level. Both hardware and software descriptions describe what is going on inside the computer, each consistent in their own way but at totally different conceptual levels.

The tension between reductionism and holism has been most convincingly portrayed by Douglas Hofstadter in his monumental work *Gödel, Escher, Bach*. His dazzling 'Ant fugue' lucidly exposes the pitfalls of level confusion by examining the fortunes of an ant colony. Ants possess an elaborate and highly organized social structure, based on the division of labour and collective responsibility. Although each individual ant has a very limited repertoire of behaviour, perhaps inferior to some modern microprocessor machines, nevertheless the colony as a whole displays a remarkable level of purpose and intelligence. The construction of the colonial home involves vast and sophisticated engineering projects. Clearly no individual ant carries a mental conception of the grand design. Each ant is simply an automaton programmed to execute a simple set of operations. This is analogous to the hardware level of description. Now consider the colony as a whole, and a complex pattern emerges. At this holistic level (corresponding to the software description in computing) emergent features, such as purposive behaviour and organization, are apparent. Collectively, a pattern emerges. Hofstadter argues that these two levels of description are not antagonistic. He denounces the question, 'Should the world be understood via holism, or via reductionism?', as invalid. It all depends what you want to know. Hofstadter points out that this perspective has long been appreciated in the East, finding expression in the cryptic oriental philosophy of Zen.

Although we are accustomed to thinking of individual ants as primary organisms, there is a sense in which the colony as a whole is also an organism. Indeed, our own bodies are also colonies, consisting of billions of individual cells co-operating in collective organization. Their association is somewhat tighter than ants in a colony, but the same basic principles of division of labour and collective responsibility are apparent. However, the crucial point to appreciate is that just as there exist emergent holistic features in an ant colony, so there are such features in a cell colony. To say that an ant colony is nothing but a collection of ants is to overlook the reality of colonial behaviour. It is as absurd as saying that computer programs are not real, they are nothing but electrical pulses. Similarly, to say that a human being is nothing but a collection of cells, which are themselves nothing but bits of DNA and so forth, which in turn are nothing but strings of atoms and therefore conclude that life has no significance, is muddle-headed nonsense. Life is a holistic phenomenon.

An appreciation of the holistic nature of life enables the old idea of a life-force to be comfortably abandoned, for that too is based on level-confusion. The idea that some magical quality must be bestowed upon inanimate matter to 'bring it alive' is as misguided as to suppose that electrical switches must be given 'computing power' or ants endowed with a 'colonial spirit', before these systems can function collectively. If it were possible to construct artificially a complete bacterium by assembling the individual atoms in the appropriate pattern, there is no doubt that it would be every bit as alive as any 'natural' bacterium.

Physicists have long since abandoned a purely reductionist approach to the physical world. This is especially true in the quantum theory, where a holistic view of the act of measurement is fundamental to the meaningful interpretation of the theory (see in Chapter 8). However, it is only in recent years that holistic philosophy has begun to have a more general impact on physical science. This changing trend has also been followed in some medical circles, with emphasis on treating 'the whole patient', and among psychologists and sociologists. Holistic science is rapidly developing into something of a cult, partly perhaps because it is in tune with oriental philosophy and mysticism. It is a shifting mood well-captured in Fritjov Capra's *The Tao of Physics* and Zukav's *The Dancing Wu Li Masters*, which exploit parallels between modern physics and traditional Eastern holistic concepts like 'oneness' of the spirit and universal destiny.

Once one has come to accept that a holistic perspective removes the need for a life-force, the question immediately arises of whether science, and in particular physics, can ever hope to provide a description of holistic phenomena, including life. One attempt to develop a wide-ranging holistic physics is made by David Bohm in *Wholeness and Implicate Order*. In dealing with biological systems, Bohm remarks: 'Life itself has to be regarded as belonging in some sense to a totality.'[2] He goes on to argue that life is somehow 'enfolded' in the whole system, including the unquestionably inanimate parts such as the air we breathe, the molecules of which may one day be incorporated in our bodies.

In fact, physics developed to cope with holistic phenomena a century ago, with the arrival of the subject of thermodynamics, through the work of James Clerk Maxwell and Ludwig Boltzmann, who attempted to deduce thermodynamic properties from the statistical properties of vast collections of molecules. Thermodynamics is a subject of central significance to life, and one which can often cast biological processes in a paradoxical light.

The paradox concerns the very essence of living things, which is *order*. As we have seen, the second law of thermodynamics, which regulates changes in order, requires disorder always to increase. But the development of life is a classic example of order on the increase. As living systems have evolved during the Earth's history into ever more complex and elaborate forms, so the level of order has risen. How can this be reconciled with the second law? Is this clash evidence that a divine agent is at work, imposing order (miraculously) on the development of biological organisms?

Closer inspection reveals that there need be no contradiction at all between biology and the second law. The latter refers always to the *total* system. It is possible for order to accumulate in one place at the price of entropy generated elsewhere. Now an essential feature of living systems is that they are 'open' to their surroundings: they are not completely sealed off or self-contained in any way. They can only survive by exchanging energy and material with their environment. When a proper entropy balance sheet is drawn up one finds that the growth of order in an organism is paid for by entropy in the wider environment. In all cases there is a net entropy increase. In fact, there are also many examples of the accumulation of order in inanimate systems. The growth of a crystal from a featureless liquid represents a local increase in order, but careful examination shows that there is a compensatory production of heat, which drives up the entropy of the surrounding material.

There is a popular belief that living things require energy, but that is not quite correct. Physics tells us that energy is conserved — it cannot be created or destroyed. When a person metabolizes food, some energy is released in his body which then dissipates into the surroundings as heat, or work performed by activity. The total energy content of a person's body remains more or less unchanged. What happens is that there is a *flow* of energy through the body. This flow is driven by the orderliness, or negative entropy, of the energy consumed. The crucial ingredient for maintaining life is, then, negative entropy. The great quantum physicist Erwin Schrödinger, in his book *What is life*?, expresses it thus:

> An organism (has an) astonishing gift of concentrating a 'stream of order' on itself and thus escaping decay into atomic chaos — of 'drinking orderliness' from a suitable environment.[3]

Being convinced that life does not defy the basic laws of physics is not, of course, the same as saying that the laws of physics explain life, only that they do not contradict it. Few physicists would claim that, given complete knowledge of the laws of atomic and molecular processes, it would ever be possible to deduce from these laws alone that life could exist. But that is not to fall into the trap of requiring 'life-force':

> An engineer, familiar with heat engines only, will, after inspecting the construction of a dynamo, be prepared to find it working along principles which he does not understand . . . The difference in construction is enough to prepare him for an entirely different way of functioning. He will not suspect that the dynamo is driven by a ghost because it is set spinning by the turn of a switch, without furnace and steam.[4]

Similarly, living organisms may operate by employing physical laws and processes that are not yet understood, even though the physics of the individual components themselves — the atoms and molecules — may be. To repeat, the collective behaviour may not be comprehensible in terms of its constituent parts. Assuming, then, that living and non-living matter obey the same laws of physics, the mystery is how a single set of laws can produce such fundamentally different behaviour. It is as though matter can branch into two pathways, one — the living — evolving towards progressively more ordered states, and the other — the lifeless — becoming more and more disordered under the impact of the second law of thermodynamics. Yet in both cases the basic constituents — the atoms — are identical.

In recent years, some progress has been made in uncovering the principles that control the appearance of collective order. The 'miracle' of life can seem less mysterious by the study of inanimate systems that also manage to achieve spontaneous self-organization. Many examples are known. To take a simple case, if a horizontal layer of liquid is heated from below, a critical temperature is reached after which the liquid organizes itself into a regular pattern of convective cells where large numbers of molecules move coherently in a recognizable pattern of flow.

The study of fluids is replete with examples of the onset of order when the system is forced away from thermodynamic equilibrium. One case concerns the appearances of vortices in fluid flow. On the Earth this produces circulation patterns in the atmosphere leading to tornados and other atmospheric disturbances. On Jupiter, similar processes cause the appearance of the characteristic elaborate and beautiful surface features.

Very striking illustrations of the spontaneous growth of order are provided by certain chemical reactions. In the so-called Belousov-Zhabotinski reaction, a chemical mix in a test tube can striate into horizontal bands, while in a shallow dish marvellous spiral forms appear. Organized chemical behaviour is frequently observed in organic (but non-living) substances under certain conditions, in many cases involving highly complex chains of reactions which incorporate some element of 'feedback' and 'catalysis'.

9 The flow of a liquid over a thin wire produces the elaborate vortices depicted here, reminiscent of surface features on the planet Jupiter. (Reproduced by courtesy of Dr. David Tritton.)

A systematic study of self-organizing systems has been carried out by the Nobel prizewinning chemist Ilya Prigogine, and his large research group at the University of Brussels. Mention should also be made of the pioneering work of Manfred Eigen. Prigogine aims not only to discover the mechanisms of self-organization but to provide a rigorous mathematical treatment for their description. In many cases the equations that describe simple behaviour patterns in advanced biological systems are the same as those that apply to inorganic chemical reactions. Prigogine believes that the principles that hold the secret of life may be manifested in these simpler examples of fluid motions or chemical mixtures. The linking feature in all his examples is that the systems concerned are driven far from thermodynamic equilibrium, whereupon they become unstable and spontaneously organize themselves on a large scale. He uses the term 'dissipative structures' to describe this organization: 'The occurrence of dissipative structures generally requires that the system's size exceed some critical value . . . (and) involve *long-range order* through which the system acts *as a whole*.'

There is no doubt that Prigogine's work has advanced greatly our understanding of far-from-equilibrium physical structures, and helped us to recognize patterns in inanimate systems that are reminiscent of living organisms. It would be foolish, however, to read too much into these results. Common behaviour does not mean common explanation. It may be that the ring shape of a benzene molecule is reminiscent of children playing ring-a-ring o' roses, but the comparison could hardly be invoked as an explanation of human behaviour.

What the study of self-organizing systems does demonstrate, however, is that the complex order of biological systems can very plausibly be attributed to highly non-equilibrium physical process, as yet only dimly perceived, but without the need for a life-force or divine spark.

Many religious people are prepared to concede that, once life had arisen on the Earth, its subsequent propagation and development can be satisfactorily explained by the laws of physics and chemistry combined with Darwin's theory of evolution. Reproduction, for example, when the DNA spiral duplicates itself chemically, seems a straightforward if complex mechanistic process. But what of the *origin* of life?

The origin of life remains one of the great scientific mysteries. The central conundrum is the threshold problem. Only when organic molecules achieve a certain very high level of complexity can they be considered as 'living', in the sense that they encode a huge amount of information in a stable form and not only display the capability of storing the blueprint for replication but also the means to implement that replication. The problem is to understand how this threshold could have been crossed by ordinary physical and chemical processes without the help of some supernatural agency.

The Earth is approximately 4½ billion years old. Traces of developed life exist in the fossil record back to at least 3½ billion years, and presumably some form of primitive life existed before this. In geological terms, then, life was quick to establish itself on our newly-cooled planet once the birth trauma of the solar system had subsided. This suggests that whatever mechanisms were responsible for the generation of life, they were quite efficient, an observation that has prompted some scientists to conclude that life is an almost inevitable result given the right physical and chemical conditions.

The favoured scenario for the origin of life is the 'primeval soup'. The primitive Earth, with its abundant supply of water, enriched by simple organic compounds formed from chemical reactions in the atmosphere, would have possessed innumerable ponds and lakes in which a vast range of chemical processes would have taken place. Over millions of years, molecules of greater and greater complexity would form until, with the threshold crossed, life itself would have arisen purely from the random self-organization of complex organic molecules.

Support for this scenario came with the celebrated Miller–Urey experiment in 1953. Stanley Miller and Harold Urey, of the University of Chicago, attempted to simulate the conditions believed to have prevailed on the primeval Earth — an atmosphere of methane, ammonia and hydrogen, a pool of water, and a thunderstorm (mimicked by an electric discharge). After a few days, the experimenters found their 'pool' of water had turned a red colour and contained many of the chemical compounds that are important in life today, such as amino acids.

Encouraging though these results are, there is no reason whatever to suppose that, left to itself, such a soup would spontaneously generate life, even after millions of years, merely by exploring every combination of chemical arrangements. Simple statistics soon reveal that the probability of the spontaneous assembly of DNA — the complex molecule that carries the genetic code — as a result of random concatenations of the soup molecules is ludicrously — almost unthinkably — small. There are so many combinations of molecules possible that the chance of the right one cropping up by blind chance is virtually zero.

However, Prigogine's work demonstrates that many systems spontaneously organize themselves if they are forced away from thermodynamic equilibrium. So rather than simply sitting there, shuffling away, the primeval soup could have been driven into a sequence of ever more complex self-organizing reactions by some external influence that upset the thermodynamic equilibrium. This influence could simply have been the sun, whose powerful flux of radiation produces the disequilibrium (negative entropy) that drives the Earth's biosphere today. Or it might have been something else; nobody knows. The end product of this sequence could have been DNA.

In summary, it is not hard to envisage a prebiotic soup containing all the necessary ingredients of biology, driven by outside disturbance into interlocking self-organizing, self-reinforcing 'feedback' loops, thereby concentrating the order and fantastically increasing the odds in favour of crossing the life threshold. But it would be wrong to suppose that we have anything like an understanding of the steps intermediate between the Miller–Urey experiment and full-blown replicating molecules. The origin of life remains a mystery, and contentious even among scientists. Indeed, Francis Crick, whose unravelling of the molecular structure of DNA in the early 1950s has been described as the discovery of the century, himself remains cautious:

> It is impossible for us to decide whether the origin of life here was a very rare event or one almost certain to have occurred . . . it seems almost impossible to give *any* numerical value to the probability of what seems a rather unlikely sequence of events.[5]

Nevertheless, a lack of understanding does not imply a miracle, and future discoveries could supply a lot of the missing details.

Even if further work suggests that a natural origin of life would imply a fantastic accident, those who believe in an infinite universe, containing an infinity of planets, need have no fear of statistics. In an infinite universe, *anything* that is possible *must* happen somewhere by pure chance. Obviously, *we* will find ourselves just where that fantastic happening has occurred.

Does the study of life — its origin and function — yield any evidence for the existence of God? We have seen that modern scientists regard life as a mechanism, and can find no real evidence of a life-force or non-material quality. The origin of life is not at all understood, though the emerging study of self-organizing systems renders a mechanistic version of biogenesis plausible to some. The remarkable ability of life to concentrate negative entropy does not, after all, lead to a violation of the second law of thermodynamics, and while the physical laws that control and direct biological functions are still only glimpsed there is no evidence that living systems actually contradict known physics and chemistry.

None of this, of course, rules out a creative God, but it does suggest that divine action may be no more necessary for biology than it is for, say, producing the rings of Saturn or the surface features of Jupiter. We either see the evidence of God everywhere, or nowhere. Life is not, it would seem, exceptionally different from other complex organized structures, except perhaps in degree. Our ignorance of the origin of life leaves plenty of scope for divine explanations, but that is a purely negative attitude, invoking 'the God-of-the-gaps' only to risk retreat at a later date in the face of scientific advance. Instead, let us regard life, not as an isolated miracle in an otherwise clockwork universe, but as an integral part of the cosmic miracle.

The general belief among scientists that life is a natural, if improbable state of matter, has encouraged speculation about the existence of alien life elsewhere in the universe. This is, of course, a contentious subject, and no attempt will be made to review it here. To date, no positive evidence exists for extra-terrestrial biology, although it has been claimed by some that the Mars Viking probe did suggest a possible biochemical reaction in one of its experiments. Nevertheless, there are probably many billions of planets in our galaxy alone, and some scientists are convinced that the universe is teeming with life. Indeed, both Hoyle and Crick have speculated that Earthlife may have come from space originally.

The possibility of alien life raises the prospect of creatures with a considerably greater intelligence than humans. Because the Earth is less than half the age of the universe, there could exist planets on which intelligent creatures evolved billions of years ago. Their intellect and technology might be unimaginably superior to our own. Beings of such advanced capabilities might well have gained control of large regions of the universe, though we can perceive no evidence of their activity.

The existence of extra-terrestrial intelligences would have a profound impact on religion, shattering completely the traditional perspective of God's special relationship with man. The difficulties are particularly acute for Christianity, which postulates that Jesus Christ was God incarnate whose mission was to provide salvation for man on Earth. The prospect of a host of 'alien Christs' systematically visiting every inhabited planet in the physical form of the local creatures has a rather absurd aspect. Yet how otherwise are the aliens to be saved?

In this space-age era, when so many people apparently accept the reality of UFOs, remarkably little attention has been given to the 'alien dimension' by the world's principal religions. According to Ernan McMullin, one of the few theologians currently to address this issue, 'religion which is unable to find a place for extraterrestrial persons in its view of the relations between God and the universe could find it increasingly difficult to command man's assent in times to come'.[6] It would be interesting to know what an alien theologian would have to say on the matter.

In our search for God, the existence of life, whether it can be explained naturally or requires miraculous intervention, provides strong evidence for some sort of purpose in the universe. But life as such is only one stage in the hierarchy of complexity. The importance of life is that it is a stepping stone to, and a vehicle for, mind, and it is to that subject that we now turn.

6. Mind and soul

> 'I think, therefore, I am.'
>
> René Descartes

> 'I simply believe that some part of the human Self or Soul is not subject to the laws of space and time.'
>
> Carl Gustav Jung

Whatever their differences of opinion about the nature of God, I know of no religion that does not teach that God is a mind. In the Christian religion God is omniscient — infinitely knowledgable. He is also infinitely free to act as he wishes. There can be no mind greater than God's, for God is the supreme being.

But what *is* mind?

This burning question has long been debated by theologians and philosophers. Today, however, the study of mind also comes within the province of science, through psychology and psychoanalysis, and more recently in brain research, computing and so-called 'artificial intelligence'. Some of these new developments have cast a wholly different light on the age-old enigma of the mind and its relation to the material world. The consequences for religion are profound. The only minds of which we have direct experience are those associated with brains (and arguably computers). Yet nobody seriously suggests that God, or departed souls, have a brain. Does the notion of a disembodied mind, let alone a mind completely decoupled from the physical universe, make any sense? In this chapter and the next we shall examine the topics of consciousness, the self and the soul, and ask whether mind can survive bodily death.

It is helpful to begin by drawing a clear distinction between the mental and physical worlds. The physical world is populated by material objects that occupy locations in space and have qualities like extension, mass, electric charge and so on. These objects are not inert, but move about, change and evolve in accordance with dynamical laws, the study of which forms a branch of physics. The physical world is (at least to a large extent) a public world, accessible by observation to everybody.

In contrast the mental world is populated not by material objects but by thoughts. Thoughts are obviously not located in space, but seem to occupy a universe of their own which is, moreover, a private universe, inaccessible to other observers. Thoughts can change, evolve, interact and otherwise behave kinetically in a variety of ways, the study of which forms a branch of psychology.

So far none of this appears controversial. Problems arise, however, when the physical and mental worlds interact. Our universe of thoughts is not isolated from the physical universe around us, but strongly coupled to it. Through our senses our minds receive a constant stream of information which proceeds to generate mental activity, either by stimulating the appearance of new thoughts or reshaping existing ones. If, when reading the sentence, you hear a loud bang from outside, the thought 'a tile has dropped from the roof' or perhaps 'a car has backfired' will intrude into your deliberations. The physical world therefore, acts as the source of new thoughts and has the effect of rearranging the mental world.

Conversely the mental world acts on the physical world through the phenomenon of volition. You decide to investigate the bang, and your legs move, the book is put down, doors open. The thoughts in your mind trigger physical activity via the intermediary of your body which then rearranges material objects in your environment. Indeed, nearly everything we ordinarily see in our environment is the result of mental activity realized through physical operations. Houses, roads, wheat fields, windmills, all originated with some intellectual activity involving planning, and decisions being converted into 'concrete reality'.

Though this all may seem obvious, there are already some disturbing features creeping in. What is the mechanism whereby matter acts on mind and, worse still, mind on matter?

Let us trace how a particular thought is 'implanted' in the mind by an external stimulus — the loud noise, for example. The sound waves impinge on the ear drum and set it into motion. The motion is transmitted through three delicate bones to the cochlea, whereupon a membrane receives the vibration and imparts it to a fluid inside the inner ear. The fluid in turn disturbs some sensitive filaments which generate electric impulses. The impulses travel along the auditory nerve pathways to the brain, where the electrical signal encounters a complex electrochemical network and the sensation of sound is registered. But how? How does the straightforward, if complex, chain of physical interactions suddenly promote a mental event — the *sensation* of sound? What is it about that particular electrochemical pattern in the brain that makes you actually *hear* something, and thereby trigger a sequence of thoughts?

Still more paradoxical is the response. You decide to investigate the sound. Your legs move — how? Brain cells fire, messages buzz along nerves, muscles tense; you move.

How would a physicist view this activity in your brain? In the first instance as processes in a complex electrical circuit, with input and output connections represented by the various nerve pathways to the sense organs and muscles. Being thoroughly familiar with the laws of electrical circuitry, the physicist might suppose that, if he could obtain a comprehensive knowledge of the electrical condition of your brain (a complete wiring diagram and detailed monitoring of the input signals) then by a stupendous computation he would be able to predict accurately the output signals of this electrical network and thereby infer what you will do next. Will you investigate the noise or not? The electrical signals will tell him.

Now nobody would suppose for one moment that such a prediction could ever be achieved. The point is, that viewed as a tangled mass of electrical circuitry, the brain seems to be completely deterministic, and therefore, in principle at least, predictable. Nerve cells fire to command your legs to move because the pattern of currents in the circuitry has a certain form. A different pattern would fail to trigger the cells and you would remain reading.

The paradox here is that these seemingly down-to-Earth physical events involving ordinary electrical impulses are paralleled by mental events: 'What is that sound? Has something broken? Should I investigate? Yes' — and the brain cells activate. But although the mental description thus far is consistent with the physical, there is a crucial element that does not tie in; namely, the fact that you *decide* to investigate the noise. The motion of the legs, setting aside of the book, and so on is the result of a conscious act of volition, a choice. Where is there room in the deterministic predictive laws of electrical circuitry for *free will*?

One answer is to view the mind as rather like the operator in control of a complicated machine. Just as a power station operator can push various buttons and light up a city, so can the mind fire the relevant brain cells (neurons) to activate the body in accordance with its decisions. But how does the conscious decision to investigate a noise *cause* the relevant brain cells to fire? What of the laws of electrical circuitry that are supposed to already determine the output signals? Are these laws violated? Can the mind somehow reach into the physical world of electrons and atoms, brain cells and nerves, and create electrical forces? Does mind really act on matter in defiance of the fundamental principles of physics? Are there, indeed, two causes of movement in the material world: one due to ordinary physical processes and the other due to mental processes?

The puzzling issue of free will and the mechanism of interaction of mind on matter will be dealt with more fully in Chapter 10. However, our problems do not rest there. We still have not discovered what consciousness is and how it arises. Are chimpanzees conscious? Dogs? Rats? Spiders? Worms? Bacteria? Computers? Is a human foetus conscious at eight months? One month? One second? Few people would answer yes to all these. So does consciousness grow gradually, is it a quality that can be quantified in some way, so that on a scale of 100 for an adult human we may assign, say, 90 for a chimpanzee, 50 for a dog, 5 for a rat, 2 for a five-month-old foetus, 0.1 for a spider and so on? Or is there a 'threshold of development' at which consciousness abruptly blossoms forth like a fuel that suddenly ignites at a critical temperature?

How can we recognize consciousness when we see it? Each of us directly experiences our own consciousness but, being located in a private, non-physical universe of thoughts and sensations it is not possible for our consciousness to be observed by anyone else. Instead, one can only infer consciousness in others through their behaviour and through communication with them via the physical universe. Jones may tell Smith that he, Jones, is conscious and Smith, observing that Jones seems a normal sort of fellow and is conducting his dialogue in a coherent way, believes him. If Jones were mute, or only spoke an obscure dialect of Eskimo, Smith would still feel confident in drawing the same conclusion through observation of Jones's conduct, with special attention to his response to stimuli, execution of complex tasks and so on.

In the case of a dog, we are on shakier ground. Dog-human communication is minimal and can be ambiguous, and much dog-behaviour seems mindless, instinctive. Yet few dog-owners would be prepared to deny that their pets are conscious and have minds, albeit less developed (in some obscure sense) than humans. But when it comes to lower creatures — spiders for example — it would be very hard to make a case that they have minds. True, they still display behaviour, but it is easy to be convinced that it is automatic — programmed by instinct.

In considering this downward progression, it is easy to be persuaded that there is an asymmetry between the way in which the active and passive aspects of mind peter out. To be conscious in the sense of registering sense-data somehow seems less accomplished than the ability to plan, decide and act. A new-born baby undoubtedly experiences sensations resulting from bodily stimuli, but is almost entirely passive in this awareness. Perhaps spiders likewise know what's going on around them but have an extremely limited capacity to respond by anything other than through a reflex action. It is often said that the ability to assess situations, plan and act accordingly, is uniquely human. That is surely a fallacy (particularly if extraterrestrial intelligent life exists). However, it may be that these more active qualities of the mind have to do not merely with awareness, but *self-awareness* (a topic discussed in the next chapter). It could be that the concept of the self is not well developed in animals.

The rapid development of powerful electronic computers has directed attention as never before to the mechanisms that underlie human thinking capabilities, and has led to some searching analyses of the relation between mind and brain. At the centre of this study is the simple yet loaded question: Can machines think?

This is not the place to review the vast literature and multiplicity of opinions about so-called 'artificial intelligence'. All experts are at least agreed that, at this time, even the most advanced computers fail to resemble the human mind in operation. As is well known, computers can usually outperform humans in arithmetic, filing and chess playing, but they still under-achieve in the composition of music and poetry. This disparity has less to do with the structural hardware of computers than the way in which they are programmed (the software). Most computers are designed to perform rather specific low-level tasks (such as huge amounts of simple arithmetic), where speed and accuracy are the overriding criteria. A computer which makes mistakes, sulks, has 'off' days or behaves in an otherwise erratic manner is of little use to most operators, though the possession of such irrational characteristics might enable it to more closely approximate human intelligence. Of course, nobody has the slightest idea how to program a computer with such human qualities, or indeed whether such a possibility exists. Nor is very much known about the operation of the human brain in this regard.

In spite of current technological limitations, the question of whether (in principle at least) machines can have 'minds' is a burning one. Anyone who has had the experience of using a powerful computer will have soon learned that, in a limited sense, it can communicate with its operator in a quasi-human fashion. Modern 'interactive' techniques enable a sophisticated dialogue to take place, on a question and answer basis, between man and machine, though the range of conversation is severely limited.

I have argued that the existence of other minds than our own can only be deduced by analogy. If one asks the question: 'How do I know that Smith has a mind?' the answer can only be: 'I have a mind, Smith behaves as I do, talks as I do, professes to have a mind, as I do, so I conclude he has a mind as I have.' But this reasoning could equally well apply to a machine as a human being. As you can never occupy the mind of another human being and experience their consciousness at first hand (and even if you could the occupied person would no longer be him, but you), any assumption about the existence of other minds is necessarily an act of faith. So the answer to the question 'Can machines think?' must be that one has no reason to rank men above machines on grounds of performance (in certain intellectual tasks) which is the only external criterion by which one can assess the machine's 'internal' experiences. If a machine could be made to respond in the same way as a human being to all external influences then there would be no observable grounds for claiming that the machine did not think, or did not have a consciousness. Moreover, if we are willing to concede that dogs think, or that spiders or ants possess some rudimentary consciousness, then even presently available computers could be regarded as conscious in that limited sense.

In 1950, the mathematician Alan Turing addressed the question 'Can machines think?' in an article entitled 'Computing Machinery and Intelligence' in the journal *Mind*. He suggested a simple test that would reveal the answer. Turing called it the 'imitation game'. The idea is that a man goes into one room, and a woman into another. An interrogator communicates with them via a teletype contraption and tries, by using a question and answer sequence, to decide which respondent is male and which female. The man and woman are asked to try and persuade the interrogator that each is the woman. Thus, the man must be a knowledgable and accomplished liar. Turing's machine intelligence test now consists of replacing the man by the machine in this game. If it succeeds in fooling the interrogator that it is a woman, Turing maintains that the machine really does think.

A number of arguments have been deployed against the claim that such full-blown artificial intelligence is possible. One line of reasoning is that computers, locked as they are in strictly rational, logical modes of operation, are inevitably cold, calculating, heartless, mindless, soulless, unemotional automata. Being purely automatic in operation, they will achieve only what has already been programmed into them by their human operators. No computer can take off and become a self-motivated creative individual, able to love, laugh, cry or exercise free will. It is no less a slave to its controllers than a motor car.

The trouble with this argument is that it can backfire. At the neural (brain cell) level, the human brain is equally mechanical and subject to rational principles, yet this does not prevent us from experiencing feelings of indecision, confusion, happiness, boredom and irrationality.

The principle religious objection to the idea of artificial intelligence is that machines do not have souls. The concept of the soul, however, is hopelessly vague. Early ideas were inextricably bound up with the concept of a life-force — some vital, animating influence. The Bible, especially the Old Testament, has very little to say on the subject, which seems to owe its origin more to the Greek scholastic tradition, under the influence of philosophers such as Plato. Early biblical references present the soul as synonymous with breath or life, but the concept sharpens somewhat in the New Testament, where the soul comes to be identified with the self and takes on the features that we might today call the mind. Indeed, the use of the word soul has declined in the modern era, and is now confined mainly to theological circles. Even the Catholic Encyclopedia settles for a definition of the soul as the 'source of thought activity'.[1] The relation between soul and mind has therefore been rather vague, and they will be used interchangably in what follows.

Central to religious doctrine is the idea that the soul (or mind) is a *thing*, and a sharp distinction must be drawn between body and soul. This so-called dualist theory of the mind (or soul) was developed by Descartes and has been widely incorporated in Christian thinking. It also comes closest to the belief of the ordinary man. Indeed, so ingrained in our culture and language are the ideas of dualism that Gilbert Ryle in his book *The Concept of Mind* calls it 'the official doctrine'.

What are the features of the dualist theory of the mind? The 'official doctrine' goes something like this. The human being consists of two distinct, separate kinds of thing: the body and the soul, or mind. The body acts as a sort of host or receptacle for the mind, or perhaps even as a prison from which liberation may be sought through spiritual advancement or death. The mind is coupled to the body through the brain, which it uses (via the bodily senses) to acquire and store information about the world. It also uses the brain as a means to exercise its volitions, by acting on the world in the fashion described earlier in this chapter. However, the mind (or soul) is not located inside the brain, or any other part of the body; or indeed anywhere in space at all. (I am discounting here the 'unofficial' doctrine of some mystics and spiritualists who claim to witness some sort of aetheric body or soul in close spatial association with the physical body.)

An important feature of this picture is that the mind is a thing; perhaps even more specifically, a substance. Not a physical substance, but a tenuous, elusive, aetherial sort of substance, the stuff that thoughts and dreams are made of, free and independent of ordinary ponderous matter.

Descartes's conception of body and soul is summarized by R.J. Hirst as follows:

> The essential notions seem to be: first that there are two distinct orders of being or substances, the mental and the material. Mind or mental substance is neither perceptible by the senses nor extended in space; it is intelligent and purposive and its essential characteristic is thought, or rather consciousness.[2]

Ryle expresses it thus:

> Though the human body is an engine, it is not quite an ordinary engine, since some of its workings are governed by another engine inside it — this interior governor-engine being one of a very special sort. It is invisible, inaudible and it has no size or weight. It cannot be taken to bits and the laws it obeys are not those known to ordinary engineers.[3]

Ryle dubs this interior governor 'the ghost in the machine'.

The soul's insubstantial quality would appear to be necessary for two reasons. First, we do not see souls or detect their physical presence in any direct way, nor are they revealed during brain surgery. Secondly, the world of matter must comply with the laws of physics which, on the macroscopic level (i.e. ignoring quantum effects) are deterministic and mechanical, and hence incompatible with free will — a fundamental attribute of the soul. (The reasoning is mistaken, as we shall see in due course.) But these arguments only tell us what the soul is not, not what it is. We get the suspicion that the idea of a soul or mind as a *thing* has been floated out of nowhere, and given a spurious and illusory impression of reality simply by attaching meaningless words to it. The mind is not mechanical, so it is 'non-mechanical', as though this adjective conveys some sort of meaning for us. According to Ryle, 'Minds are not bits of clockwork, they are just bits of non-clockwork'.[4]

Difficulties also lie in store when we try to understand where, precisely, the soul is located. If it is not to be found in space, where is it? (It is interesting to note, however, that Descartes believed the small pineal gland in the brain was the seat of the soul, or at least was the structure that provides the elusive physical link between mind and brain.) Can the new physics, with its weird concepts of spacewarps and higher dimensions provide a suitable location?

We have seen how physicists think of space and time as a sort of four-dimensional sheet (or perhaps balloon) with the possibility of other disconnected sheets. Could the soul reside in one of these other universes? Alternatively, spacetime may be envisaged as enfolded by, or embedded in, a higher dimensional space, much as a two-dimensional surface or sheet is embedded in three-dimensional space. Might not the soul inhabit a location in this higher dimensional space which is still (geometrically speaking) close to our physical spacetime, but not actually in it? From this higher dimensional vantage point the soul could 'lock on' to the body of an individual in spacetime, without itself being part of spacetime.

For those who wish to believe that departed souls travel to Heaven, a more complicated arrangement would be necessary, for presumably the place which souls inhabit during the Earthly life of an individual is not the same as Paradise. If such ideas strain credulity as much as geometrical intuition, it is surely because of the dubious assumption that the soul has a location. To say that the soul occupies a *place* means that it exists in some sort of space, either the one we ordinarily perceive, or some other. In that case one may then ask questions about the size, shape, orientation and motion of the soul, all concepts that are totally inappropriate to something composed of thoughts rather than materials.

But the fund of ideas from modern physics is not yet exhausted. As explained in Chapter 3, some physicists now think of space and time as derived, rather than primitive concepts. They believe that spacetime is built up out of subunits (not places or moments, but abstract entities) that would also embody quantum features. It could then be that the physical universe extends beyond (in a figurative sense) what we ordinarily call spacetime; that only a fraction of these subunits have come together in an organized way to produce spacetime, leaving 'elsewhere' a sort of ocean of disconnected bits. Could this ocean be the realm of the soul? If so, the soul would not occupy a place, because the subunits would not be assembled into places, so concepts like extension or orientation would be meaningless. Indeed, even topological concepts such as inside, outside, between, connected and disconnected, might be undefined. I leave the question open.

Further problems crop up when one turns to the question of time. A soul is not in space, but is it in time? Presumably the answer is yes. If the soul is the source of our perceptions, then this must include our perception of time. Moreover, many recognizably human mental processes are explicitly time-dependent: planning, hoping, regretting, anticipating, for instance.

There would be grave logical difficulties with a timeless soul. What meaning do we then attach to the soul's existence *after* death, if the before–after relation is transcended by souls? What about the soul's situation before the birth of the body? This issue is tackled by the Catholic Encyclopedia with a rare touch of humour:

> The notion that God has a supply of souls that are not any body's in particular until He infuses them into human embryos is entirely unwarranted by any evidence . . . The soul is created by God at the time it is infused into matter.[5]

The message is unmistakable. There are times (before birth) when the soul does not exist. Such notions clearly conflict with the idea of the soul transcending time.

The same basic temporal dilemma runs through all discussions of immortality. On the one hand is the desire for a continuation of the personality after Earthly life has ended — not merely in a frozen or timeless existence, but involving some sort of activity. Jesus spoke of 'life everlasting', which carries connotations of the unending passage of time.

On the other hand, such notions are strongly tied to our perception of time in the physical world, and do not accord well with the alleged separation of the physical and spiritual realms. The difficulty is exacerbated if one entertains the possibility (to be discussed in Chapter 15) that there may actually be an end to time: there may be no 'everlasting' anyway.

The arguments presented here, and others, have suggested to many people that the concept of the soul or mind and its immortality is at best wrong and at worst incoherent.

Several alternatives to dualism have been discussed by philosophers. At one extreme is materialism which denies the existence of mind altogether. The materialist believes that mental states and operations are nothing but physical states and operations. In the field of psychology materialism becomes what is known as behaviourism, which proclaims that all humans behave in a purely mechanical way in response to external stimuli. At the other extreme is the philosophy of idealism which asserts that it is the physical world that does not exist; everything is perception.

It seems to me that the dualist theory falls into the trap of seeking a substance (the mind) to explain what is really an abstract concept, not an object. The temptation to reduce abstract concepts to things is apparent throughout the history of science and philosophy, illustrated by discredited concepts like phlogiston, the fluid theory of heat, the luminiferous aether and the life-force. In all these cases the associated phenomena require explanation in terms of the abstract, such as energy or fields.

The fact that a concept is abstract rather than substantial does not render it somehow unreal or illusory. A person's nationality cannot be weighed or measured, it does not occupy a location inside their bodies, and yet it is a meaningful and important part of their make-up as anyone unfortunate enough to find themselves stateless knows only too well. Concepts like usefulness, organization, entropy and information do not involve 'things' in the sense of objects, but relationships between, and conditions of, objects.

The fundamental error of dualism is to treat body and soul as rather like two sides of a coin, whereas they belong to totally different categories. Ryle blames such a category mistake for all the muddle, confusion and paradox concerning the mind and its relation to the body:

> It is perfectly proper to say, in one logical tone of voice, that there exist minds and to say, in another logical tone of voice, that there exist bodies. But these expressions do not indicate two different species of existence.[6]

The statements 'there exist rocks' and 'there exist Wednesdays' are both correct, but it would be meaningless to place rocks and Wednesdays alongside each other and discuss their interrelation. Or, to use one of Ryle's analogies, it would be absurd to discuss whether there had been any discourse between the House of Commons and the British

Constitution. These institutions belong to different conceptual levels.

Ryle thus anticipates much of the 'holistic' discussion of recent years. As we saw in the previous chapter, the relation between mind and body is similar to that between an ant colony and ants, or between the plot of a novel and the letters of the alphabet. Mind and body are not two components of a duality, but two entirely different concepts drawn from different levels in a hierarchy of description. We are back to holism versus reductionism once more.

Many of the old problems of dualism fall away once it is appreciated that abstract, high-level concepts can be equally as real as the low-level structures that support them, without any mysterious extra substances or ingredients. Just as a life-force is an unnecessary addition for matter to become animate, so a soul-substance is unnecessary for matter to become conscious:

> Our world is filled with things that are neither mysterious and ghostly nor simply constructed out of the building blocks of physics. Do you believe in voices? How about haircuts? Are there such things? What are they? What, in the language of the physicist, is a hole — not an exotic black hole, but just a hole in a piece of cheese, for instance? Is it a physical thing? What is a symphony? Where in space and time does 'The Star Spangled Banner' exist? Is it nothing but some ink trails on some paper in the Library of Congress? Destroy that paper and the anthem would still exist. Latin still *exists*, but it is no longer a living language. The language of the cavepeople of France no longer exists at all. The game of bridge is less than a hundred years old. What sort of a thing is it? It is not animal, vegetable, or mineral.
>
> These things are not physical objects with mass, or a chemical composition, but they are not purely abstract objects either — objects like the number π, which is immutable and cannot be located in space and time. These things have birthplaces and histories. They can change, and things can happen to them. They can move about — much the way a species, a disease, or an epidemic can. We must not suppose that science teaches us that every *thing* anyone would ever want to take seriously is identifiable as a collection of particles moving about in space and time. Some people may think it is just common sense (or just good scientific thinking) to suppose *you* are nothing but a particular living, physical organism — a moving mound of atoms — but in fact this idea exhibits a lack of scientific imagination, not hard-headed sophistication. One doesn't have to believe in ghosts to

> believe in *selves* that have an identity that transcends any particular living body.[7]

The brain consists of billions of neurons, buzzing away, oblivious of the overall plan (like the ants in the colony discussed in the previous chapter). This is the physical, mechanical, world of electrochemical hardware. On the other hand we have thoughts, feelings, emotions, volitions and so on. This higher level, holistic, *mental* world is equally oblivious of the brain cells; we can happily think while being totally unaware of any help from our neurons. But the fact that the lower level is ruled by logic need not contradict the fact that the upper mental level can be illogical and emotional. Hofstadter has given a vivid illustration of this neural-mental complementarity:

> Say you are having a hard time making up your mind whether to order a cheeseburger or a pineappleburger. Does this imply that your neurons are also balking, having difficulty deciding whether or not to fire? Of course not. Your hamburger-confusion is a high-level state which fully depends on the efficient firing of thousands of neurons in very organized ways.[8]

To use an anology, a competently written novel will consist of a sequence of grammatical constructions conforming to rather precise logical rules of language and expression. Yet this does not prevent the characters in the novel from loving and laughing, or behaving in a completely unruly way. To claim that because the book is built out of logical word constructions obliges the story itself to comply with rigid logical principles would be absurd. It is to confuse two distinct levels of description. MacKay has also emphasized the importance of avoiding level-confusion when discussing neural versus mental activity: 'The idea that one and the same situation may need two or more accounts, each *complete* at its own logical level, may sound abstract and difficult. But as we have seen, it can be illustrated by numerous examples.' Discussing his analogy of the illuminated advertising display which is completely explicable in terms of electrical circuit theory, MacKay points out that it has a complementary description in terms of the commercial message: 'When properly disciplined, these (two descriptions) are not rivals, but complementary, in the sense that each reveals an aspect which is there to be reckoned with, but is unmentioned in the other.'

Thus, when it comes to the mind:

> The notion, popularized by writers like Teilhard de Chardin, that if men are conscious there must be some traces of consciousness in atoms, is quite without rational foundation . . . Consciousness

> is not something we expect to be forced to recognize as the end-product of an argument about the behaviour of physical particles . . .[9]

In more modern parlance, the mind is 'holistic'.

None of this, of course precludes the possibility of artificial minds, thinking machines, and so forth. It is curious that many people who readily accept that their pets have minds shudder at the thought of a computer with a mind. Perhaps it is an egocentric reaction to the threat that one day computers may have minds of greater intellectual power than our own. Or perhaps it is more subtle.

The two-level (or multi-level) description of mind and body is a great improvement on the old idea of dualism (mind and body as two distinct substances) or materialism (mind does not exist). It is a philosophy that is rapidly gaining ground with the emergence of what are known as the cognitive sciences: artificial intelligence, computing science, linguistics, cybernetics and psychology. All these fields of enquiry are concerned with systems that process information in one way or another, whether man or machine. The development of concepts and language associated with computers, such as the distinction between hardware and software, has opened up new perspectives on the nature of thought and consciousness. It has forced scientists to think more clearly than ever before about the mind.

These scientific advances have been matched by the appearance of a new philosophy of mind, closely tied to the ideas presented above, called functionalism. Functionalists recognize that the essential ingredient of mind is not the hardware — the stuff your brain is made of or the physical processes that it employs — but the software — the organization of the stuff, or the 'program'. They do not deny that the brain is a machine, and that neurons fire purely for electrical reasons — there are no mental causes of physical processes. Yet they still appeal to causal relations between mental states: very crudely, thoughts cause thoughts, notwithstanding the fact that, at the hardware level, the causal links are already forged.

That there is no incompatibility between the causal connections at the hardware and software levels is taken for granted by most computer programmers. In one breath they will say: 'The computer is simply a lot of circuitry and anything it can do is determined by the laws of electricity. Its output is an automatic consequence of its following predetermined electrical pathways.' Then they will talk about the computer solving equations, making comparisons and decisions and arriving at conclusions based on information processes, i.e. pushing ideas round. So it is possible to live with two different levels of causal description — hardware and software — without ever having to grapple with how the software acts on the hardware. The old conundrum of how the mind acts on the body is seen to be just a muddle of conceptual levels. We never ask 'How does a computer program make its circuits solve the equation?' Nor do we need to ask how thoughts trigger neurons to produce bodily responses.

What does functionalism imply for religion?

It seems to be something of a double-edged sword. On the one hand functionalism denies that mind is uniquely human, and claims that machines can also think and feel, at least in principle. It is hard to reconcile that viewpoint with the traditional notion of God endowing man with a soul. On the other hand, by liberating mind from the confines of the human body, it leaves open the question of immortality:

> The software description of the mind does not logically require neurons . . . it allows for the existence of disembodied minds . . . Functionalism does not rule out the possibility, however remote it may be, of mechanical and ethereal systems having mental states and processes.[10]

Functionalism solves at a stroke most of the traditional queries about the soul. What stuff is the soul made of? The question is as meaningless as asking what stuff citizenship or Wednesdays are made of. The soul is a holistic concept. It is not made of stuff at all.

Where is the soul located? Nowhere. To talk of the soul as being in a place is as misconceived as trying to locate the number seven, or Beethoven's fifth symphony. Such concepts are not in space at all.

What of the problems about time and the soul? Does existence in time but not space make any sense?

Here the issue is more subtle. We frequently talk about rising unemployment or changing fashions, implying the time-dependence of things that cannot be meaningfully pinned down at a distinct place. There seems to be no reason why the mind cannot evolve with time even though it is not to be found anywhere in space.

We may therefore choose to reject the belief that mind is nothing but brain cell activity, for that is to fall into the reductionist trap. Nevertheless, it seems that the existence of the mind is supported by that activity, and so the question arises of how disembodied minds can exist. To resort yet again to analogy, a novel is built out of words, but the story could equally well be stored vocally on magnetic tape, coded on punched cards or digitally on computer, for example. Can the mind survive the death of the brain by being transferred to some other mechanism or system? Clearly this would be possible in principle.

Most people, however, do not contemplate the survival of their entire personality; so much of our makeup is tied to our bodily needs and capabilities. Sexuality, for example, in the absence of a body or a need for procreation, would be ridiculous. Many would also not wish for the negative aspects of their personality — the greed, jealousy, hatred and so forth — to survive. The enduring core of mind would have to be stripped of its more obviously bodily associations and unpleasant features. But would anything then be left? What about personal identity — the *self*?

7. The self

> 'Each self is a divine creation.'
>
> Sir John Eccles

> 'My one regret in life is that I'm not someone else.'
>
> Woody Allen

What are we? Each of us has buried deep within our consciousness a strong sense of personal identity. As we grow up and develop, our opinions and tastes change, our perspective of the world shifts, new emotions surface. Yet through it all we never doubt that we are the same person. *We* have those changing experiences. The experiences happen to *us*. But what is the 'we' that has the experiences? That is the long-standing mystery of the self.

When dealing with other people we usually identify them with their bodies, and to a lesser extent their personalities, but we view ourselves quite differently. When someone refers to 'my body' it is in the sense of a possession, as in 'my house'. But when it comes to mind, that is not so much a possession as a *possessor.* My mind is not a chattel: it is *me*.

The mind is, then, regarded as the *owner* of experiences and feelings, the centre or focus of thoughts. My thoughts and my experiences belong to me; yours belong to you. In the words of the Scottish philosopher Thomas Reid:

> Whatever this self may be it is something which thinks, and deliberates, and resolves, and acts, and suffers. I am not thought, I am not action, I am not feeling; I am something that thinks, and acts and suffers.[1]

What more natural than for theologians to identify the self with the elusive mental substance or soul? Furthermore, as the soul is not located in space, it cannot be 'pulled apart' or disseminated, so the integrity of the self is assured. For it is one of the most fundamental properties of the perceived 'self' that it is indivisible and discrete. *I* am *one* individual, and *I* am quite distinct from *you*.

The concept of the mind (or soul), as we saw in the previous chapter, is, nevertheless, a notoriously difficult one and can involve paradox. The question 'What *am* I?' is not an easy one to answer. As Ryle points out: 'Gratuitous mystification begins from the moment that we start to peer around for the beings named by our pronouns.'[2] Still, the question has to be answered if one is to make any sense at all of the idea of immortality. If I am to survive death just what *is* it that I can expect to survive?

According to David Hume, the self is nothing but a collection of experiences:

> When I enter most intimately into what I call *myself* I always stumble on some particular perception or other, of heat or cold, light or shade, love or hatred, pain or pleasure. I can never catch *myself* at any time without a perception, and never can observe anything but the perception.[3]

So adopting this philosophy, the answer to the question 'What am I?' is simply 'I am my thoughts and experiences'. Yet there is a feeling of unease about this. Can thoughts exist without a *thinker*? And what is there to distinguish *your* thoughts from *my* thoughts? What, indeed, does one mean by 'my' thoughts? In fact, Hume was later to write of his first assessment: 'Upon a more strict review of the section concerning *personal identity*: I find myself involved in a labyrinth.'

It has to be conceded, however, that the concept of self is nebulous, and that experiences go a long way to shaping the quality of the self, even if they do not explain it completely away. Some aspects of the self seem to lie on the borderline of personal identity. Where are we to locate (figuratively) emotions for example? Do you *have* emotions (as you have a body) or are your emotions an integral part of *you*? It is well known that emotions are strongly influenced by physical effects, such as the chemical composition of the blood. Hormone imbalances can produce various emotional disorders. Drugs can produce or depress a variety of mental states and emotional dispositions (as any consumer of alcohol knows). More drastically, brain surgery can produce major alterations of personality. All this makes us reluctant to clothe the soul with too many of the trappings of personality. On the other hand, if *all* emotions are removed, what is left? Christians, for example, might accept shedding negative emotions, but would wish the soul to retain feelings of love and reverence. Morally neutral feelings like boredom, vigour and a sense of humour are presumably debatable.

Of greater concern is the question of memory and the whole issue of our perception of time. Our conception of ourselves is strongly rooted in our memory of past experiences. It is not at all clear that, in the absence of memory, the self would retain any meaning whatever. It might be objected that a person suffering from amnesia may still wonder 'Who am I?' but does not for a moment doubt that there exists an 'I' to whom the 'who' pertains. Still, even an amnesiac is not completely deprived of memory. He has no difficulty, for example, in knowing the use of everyday objects, such as cups and saucers, buses and beds. Furthermore, his short-term memory remains unaffected: if he decides to walk in the garden he does not a few moments later wonder what he is doing there.

If a person did lose the ability to remember his experiences of even a few seconds previously then his sense of identity would completely disintegrate. He would be unable to act or behave coherently at all. His bodily movements would not be coordinated in any conscious pattern of action. He would be totally incapable of making any sense of his perceptions, and could not even begin to order his experiences of the world about him. The whole notion of *himself* as distinct from his perceived world would be chaotic. No pattern or regularity in events would be apparent, and no concept of continuity — especially personal continuity — could be maintained.

It is thus largely through memory that we achieve a sense of personal identity, and recognize ourselves as the *same* individual from day to day. Throughout life we inhabit one body, but the body can undergo considerable changes. Its atoms are systematically replaced as a result of metabolic activity; it grows, matures, ages, and eventually dies. Our personalities also undergo major changes. Yet through this continuous metamorphosis, we believe that we are one and the same person. If we had no memory of earlier phases of our life, how could the concept 'same person' have any meaning, save in the sense of bodily continuity?

Suppose a man claimed to be a reincarnation of Napoleon. If he did not look like Napoleon the only criterion by which you could judge his claim would be that of memory. What was Napoleon's favourite colour? How did he feel before the battle of Waterloo? You would expect him to relate some specific (and preferably verifiable) information about Napoleon before taking the claim seriously. Suppose, however, that the man declared that he had lost all memory of his previous life, save only that he *was* Napoleon, what should you make of it? What would it mean for him to say 'I was Napoleon'?

'What I mean,' he would perhaps counter, 'is that, although my body and my memory, and indeed my entire personality, are now those of John Smith, the soul of John Smith is none other than that of the late Napoleon Bonaparte. I *was* Napoleon, *now* I am Smith, but it is the same *me*. Only my characteristics have changed.' But is this not jibberish? For what is to identify one person's mind from another other than their personality or their memory? To claim that there is some sort of transferable label — the soul — which is otherwise quite devoid of properties save to display some mystical registration mark, is a totally meaningless conjecture. What would we say to someone who denied its existence? Could we not invent souls for everything in this way — for plants and clouds, rocks and airplanes? 'This looks like an ordinary diesel locomotive,' one might declare, 'but in fact it contains the essence, the soul, of Stevenson's original Rocket! The design is different, the materials are different, the performance bears no resemblance to the Rocket, but it is actually the *same* locomotive with a totally new structure, appearance and design.' What is the use of such an empty assertion?

To take a more plausible example than reincarnation, suppose that a close friend were to undergo major surgery which is so comprehensive that he is physically totally unrecognizable afterwards. How would you know he was the same person? If he related to you facts about his earlier life, reminded you of small incidents and personal conversations, and generally displayed a good acquaintance with his former circumstances, you would be inclined to conclude it was indeed the same man. 'That's him all right. Nobody else could know that.' But if the surgeon also removed much of your friend's memory, or perhaps damaged it, your judgment of his identity would be far less confident. If he had no memory at all, you would have no grounds (except perhaps some residual bodily evidence) for saying that the man before you was your friend. In fact, it is not clear that someone with *no* memory can really be thought of as a person at all; he would possess none of the coherent features, such as a personality, that we would normally associate with an 'individual'. His responses would either be totally random, or pure reflex, and as such his behaviour would be little different from a rather badly programmed automaton.

The difficulty that one sees here for the dualist who believes in the survival of the soul is obvious. If the soul depends on the brain for memory storage, how can the soul remember anything after the death of the body? And if it can't remember anything, what right have we to attribute a personal identity to it? Or are we to suppose that the soul has a sort of non-material back-up memory system that functions in parallel with the brain but can equally well cope on its own?

Sometimes an attempt is made to break out of this deadlock by asserting that the soul transcends time. Just as the soul cannot be located in space, so it has no location in time. But this manoeuvre brings with it a fresh crop of difficulties, as we saw in the previous chapter.

We seem to approach closer to an understanding of the self by noting a point made by many philosophers: that human consciousness does not consist merely of awareness, but of self-awareness — we *know* that we know. In 1690 John Locke emphasized that it is 'impossible for anything to perceive without *perceiving* that he does perceive.'[4] The Oxford philosopher, J.R. Lucas, expresses this point as follows:

> In saying that a conscious being knows something, we are saying not only that he knows it, but that he knows he knows it, and that he knows he knows he knows it, and so on . . . The paradoxes of consciousness arise because a conscious being can be aware of itself, as well as of other things, and yet cannot really be construed as being divisible into parts.[5]

In similar vein A.J. Ayer has written: 'There is a temptation to think of one's self as a set of Chinese boxes, each surveying the one it immediately encloses.'[6]

There is no doubt that the quality of self-reference is a key one in unravelling the mystery of the mind. We have already encountered the importance of feedback and self-coupling in Prigogine's dissipative structures which have the capacity for self-organization, and there seems to be a natural progression from the inanimate through the animate to the conscious — a hierarchy of complexity and self-organization. But there is another hierarchy buried in this progression — a hierarchy of conceptual levels discussed in the previous chapter. Life is a holistic concept, the reductionist perspective revealing only inanimate atoms within us. Similarly mind is a holistic concept, at the next level of description. We can no more understand mind by reference to brain cells than we can understand cells by reference to their atomic constituents. It would be futile to search for intelligence or consciousness among individual brain cells — the concept is simply meaningless at that level. Clearly, then, the property of self-awareness is holistic, and cannot be traced to specific electrochemical mechanisms in the brain.

The study of self-reference has always encountered a touch of paradox, not only in the philosophical question of self-awareness but in the arts and even at the logical and mathematical level. The Greek scholar Epeminides drew attention to the problems of self-referring statements. Normally we assume that every meaningful statement must either be true or false. But consider Epeminides' proposition (which we call A) that can be paraphrased thus:

A: This statement is false.

Is A true, or false? If true, then the statement itself declares it is false; if false, the statement must be true. But A cannot be both true and false, so the question 'Is A true, or false?' has no answer.

We ran into a similar problem in the form of Russell's paradox in Chapter 3. In both cases, absurdity seems to follow from perfectly innocuous statements or concepts when they are looped around and directed at themselves. An equivalent form of A is:

A: The following statement is true. A1
The preceding statement is false. A2

In this form, each individual statement, A1 and A2, is perfectly straightforward and free of paradox, but joined together into a self-referring loop they appear to make nonsense of logic.

In his remarkable book, Hofstadter points out how 'locally' sensible concepts that loop into paradox 'globally' have received dramatic artistic representation in the work of the Dutch artist M.C. Escher. Consider his *Waterfall* for example. If we follow the path of the water around the loop, at each stage its behaviour seems perfectly normal and natural until suddenly, with a shock, we find ourselves back where we started. The entire loop, taken as a whole, is manifestly an impossibility, yet at no point on the path around the loop does anything go 'wrong'. It is purely the global, or holistic, aspect that is paradoxical. Hofstadter also finds a musical equivalent of these 'strange loops' in Bach's fugues.

Penetrating investigations of self-reference have been carried out by mathematicians and philosophers concerned with the logical foundations of mathematics. Perhaps the most startling accomplishment of this program is a result proved by the German mathematician Kurt Gödel in 1931, known as the Incompleteness Theorem, which forms the linking theme of Hofstadter's book. Gödel's theorem sprang from the attempt by mathematicians to systematize the process of reasoning in order to clarify the logical basis on which the edifice of mathematics is built. Russell's paradox, for example, arose from efforts to organize concepts in as general and non-committal a way as possible by allocating them to 'sets' — with disastrous results.

Gödel hit upon the idea of using mathematical objects to codify statements. In itself that is nothing new or sensational. Anyone who has read an enumerated contract is familiar with the practice. The novel feature which Gödel explored was the use of mathematics to codify statements about mathematics — the self-referring aspect again. Perhaps inevitably, something similar to the Epeminides' paradox emerged, but as a statement about mathematics; in fact, about good old-fashioned numbers 1,2,3 . . . Gödel demonstrated in his theorem that there always exist statements about numbers that can *never*, even in principle, be proved either true or false (like A above), on the basis of a fixed set of axioms. Axioms are the things you assume are true without proof (e.g. 1 = 1). Thus, even a mathematical system as relatively simple as the theory of numbers possesses properties that cannot be proved (or disproved) on the basis of a fixed set of assumptions, however complex and numerous those assumptions may be!

The importance of Gödel's Incompleteness Theorem is that, by mixing subject and object, it demonstrates how, even at the fundamental level of logical analysis, self-reference can produce either paradox or indecision. It has been taken to imply that one can never, even in principle, understand one's own mind completely. Hofstadter conjectures: 'Gödel's Incompleteness Theorem (has) the flavour of some ancient fairy tale which warns you that "To seek self-knowledge is to embark on a journey which . . . will always be incomplete".'[7]

Gödel's theorem has also been used to argue for the non-mechanical nature of the mind. In an essay entitled 'Minds, Machines and Gödel', Lucas asserts that human intelligence can never be attained by computers: 'Gödel's theorem seems to me to prove that Mechanism is false, that is, that minds cannot be explained as machines.' The essence of his argument is that we, as human beings, can discover mathematical truths about numbers which a computer, programmed to work within a fixed set of axioms, and therefore subject to Gödel's theorem, cannot prove for itself:

> However complicated a machine we construct it . . . will be liable to the Gödel procedure for finding a formula unprovable-in-that-system. This formula the machine will be unable to produce as being true, although a mind can see it is true. And so the machine will still not be an adequate model of the mind.[8]

No doubt many people would feel uncomfortable about basing the mind's superiority on esoteric mathematics, when it is usually qualities such as love, appreciation of beauty, humour, and so on that are cited as evidence for a non-mechanical mind, or 'soul'. In any case, Lucas's argument has been attacked on a number of grounds. For example, Hofstadter points out that, in practice, the human mind's capacity for discovering complicated mathematical truths is limited, so one could still program a computer that could successfully prove everything that a given person can ever discover about numbers. Moreover, it is easy to convince oneself that *we* are just as vulnerable as computers to Gödel incompleteness because of Epimenides-type statements; it is possible to construct logical truths about the world involving Smith that can never be proved by Smith!

As emphasized in the foregoing, consciousness, the impression of free will and the sense of personal identity all involve an element of self-reference and can have paradoxical aspects. When a person perceives something — a physical object, for instance — the observer is by definition external to the observed object, though coupled to it through some sensory mechanism. But during introspection — an observer observing himself — both subject and object coincide in a most perplexing way. It is as though the observer is both inside and outside himself.

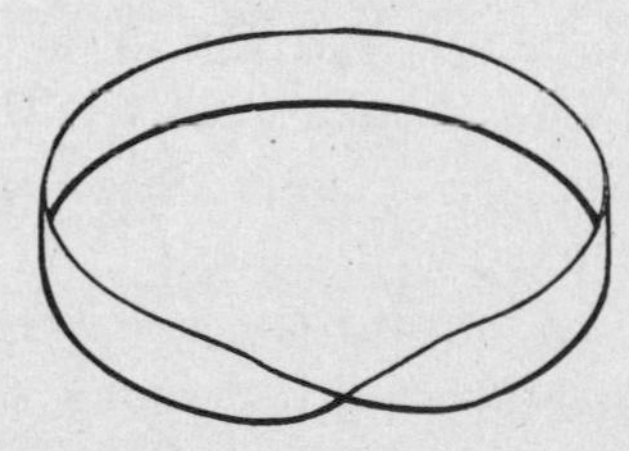

10 The famous Möbius band is made by putting a single twist in a strip and joining the ends to form a loop. Careful examination reveals that the band now has only one side and one edge.

Some intriguing representations can be given for this curious mental topology. Consider the famous Möbius band (see Fig. 10), for example. The band is constructed by making a single twist in a strip of material and then joining it into a closed loop. At any particular point on the band, there would seem to be both a front side and reverse side of the band. But if you follow a route around the loop, you will see that there is actually only one side. Locally there appears to be a division into two categories (analogous to subject and object) but a glance at the global structure shows there is only one.

Another suggestive representation of self-reference is given by Hofstadter in the language of his Strange Loops:

> My belief is that the explanations of 'emergent' phenomena in our brains — for instance, ideas, hopes, images, analogies, and finally consciousness and free will — are based on a kind of Strange Loop, an interaction between levels in which the top level reaches back down towards the bottom level and influences it, while at the same time being determined by the bottom level . . . The self comes into being the moment it has the power to reflect itself.[9]

The essential feature in all these attempts to grope towards a better understanding of the self is the convolution of hierarchical levels. The hardware of brain cells and electrochemical machinery supports the software level of thoughts, ideas, decisions, which in turn couple back to the neural level and so modify and sustain their own existence. The attempted separation of brain and mind, body and soul, is a confusion born of trying to sever these two convoluted levels (or 'Tangled Hierarchy' in Hofstadter's parlance). But it is a meaningless enterprise, for it is the very entanglement of the levels that makes you *you*.

Remarkably, modern Christian doctrine has moved a long way towards this picture of the integrated brain and mind, with its emphasis on the resurrection of the *whole man* through Christ, rather than the traditional idea of a distinct immortal soul being cast adrift from its material counterpart to carry on a disembodied existence somewhere.

However, nothing that has been said about the mind is specifically restricted to human beings. There seems to be no scientific evidence for any special divine quality in man, and no fundamental reason is apparent why an advanced electronic machine should not, in principle, enjoy similar feelings of consciousness as ourselves. This is not, of course, to say that computers have souls, but rather that the complex tangle of convoluted levels which produce what we understand as mind can arise in a variety of systems.

Yet there still remains one aspect of the self that seems to be contradicted by the low-level, deterministic description, and that is the *will*. All human beings believe that they are capable of choosing, in a limited way, between various courses of action available to them. Can such an apparent freedom to initiate actions ever be programmed into a computer?

Hofstadter argues that in principle we can. He describes the feeling we have of free will as a delicate balance between self-knowledge and self-ignorance. By incorporating an appropriate degree of self-reference into a computer programme, Hofstadter claims that it too would start to behave as though it had a will of its own. He tries to tie in the will with the Gödel-like incompleteness which inevitably arises in any system capable of monitoring its own internal activity. (The subject of free will and determinism will be explored in greater depth in Chapter 10.)

Suppose that one were persuaded by these arguments that human brains are marvellously complex electrochemical machines, and that other types of artificial mechanisms, such as computers, can be programmed for free will and human-like emotions. Does this devalue the human mind? Recall the trap of 'nothing-buttery'. To assert that the brain is a machine does not deny the reality of mind and emotion, which refers to a higher level of description (the ant colony, the plot of the novel, the picture on the jig-saw, the Beethoven symphony). To say a brain is a machine need not imply that the mind is *nothing but* the product of mechanistic processes. To claim that the deterministic nature of brain activity renders free will an illusion is as misconceived as the claim that life is an illusion because of the underlying inanimate nature of atomic processes.

A number of science fiction writers have developed the idea of machines with minds, most notably Isaac Asimov in his robotics stories, and Arthur C. Clarke in the novel *2001: A Space Odyssey*. More penetrating analyses have also been given by some writers who visualize 'mind transplants' in an attempt to clarify the definition of the self.

Consider, for example, what would happen if your brain could be removed and placed in a 'brain support system', remaining coupled to your body via some sort of radio communication network. (Of course, such a procedure is utterly beyond foreseeable technology but there is no logical reason why it could not be achieved.) Your eyes, ears and other senses remain functioning as usual. Your body can operate without impediment. In fact, nothing would seem any different (perhaps a feeling of light-headedness!), except that you could look down upon your own brain. The question is, where would *you* be? If your body takes a train journey, your experiences are those of someone on the journey, exactly as they would be if your brain were still in your skull. You would certainly *feel* as though you were on the train.[10]

Perplexity mounts if we now envisage your brain being transplanted into another body instead. Would it be correct to say that *you* had a new body, or that *it* had a new brain? Could you regard yourself as the same person, with a different body? Perhaps you could. But suppose the body were of the opposite sex, or that of an animal? Much of what makes *you*, your personality, capabilities and so forth, is tied to chemical and physical conditions of the body. And what if your memory were wiped out during the transfer? Does it then make any sense at all to regard the new individual as *you*?

Fresh problems arise when one speculates about duplication of the self. Suppose the entire information content of your brain were put on a giant computer somewhere, and your original body and brain died. Would *you* still survive — in the computer?

The idea of putting minds into computers raises the prospect of multiple duplicates of you being copied on other computers. Of course, much has already been written about 'multiple personality' mental disorders, and the cases where patients have had the connections between the left and right hemispheres of their brains severed, leading to mental states where, crudely speaking, the left hand literally does not know what the right hand is doing.

Though some of these ideas may seem fearsome, they do hold out the hope that we can make scientific sense of immortality, for they emphasize that the essential ingredient of mind is *information*. It is the pattern inside the brain, not the brain itself, that makes us what we are. Just as Beethoven's Fifth Symphony does not cease to exist when the orchestra has finished playing, so the mind may endure by transfer of the information elsewhere. We considered above how, in principle, the mind can be put on a computer, but if the mind is basically 'organized information' then the medium of expression of that information could be anything at all; it need not be a particular brain or indeed any brain. Rather than 'ghosts in machines', we are more like 'messages in circuitry' and the message itself transcends the means of its expression.

MacKay expresses the viewpoint in computer language:

> If a computer operating a given program were to catch fire and be destroyed, we would certainly say that that was the end of that particular embodiment of the program. But if we wanted that same program to run in a fresh embodiment, it would be quite unnecessary to salvage the original computer parts or even to replicate the original mechanism. Any active medium (even operations with pencil-and-paper) which gave expression to the same structure and sequence of relationships could in principle embody the very same program.[11]

This conclusion leaves open the question of whether the 'program' is re-run in another body at a later date (reincarnation), or in a system which we do not perceive as part of the physical universe (in Heaven?), or whether it is merely 'stored' in some sense (limbo?). As far as the perception of time is concerned, we shall see that it is only during the running of the program, as in the actual playing of a symphony, that any meaning can be attached to the flow of time. The existence of a program, like that of a symphony, once created, is essentially timeless.

In this chapter it has been argued that research in the cognitive sciences has tended to emphasize the similarities between mind in man and machine, with mixed implications for religion. While on the one hand these studies leave little room for the traditional idea of the soul, on the other hand they leave open the possibility of survival of the personality.

Minds, being complex, are not usually studied in the framework of physics that, as we have seen, operates best at the reductionist level on simple elementary things. However, there is one important area of the new physics into which mind has intruded at a fundamental level, much to the mystification of physicists. It is called the quantum theory, and it leads us into an Alice-in-Wonderland world that cuts right across the traditional framework of religion.

8. The quantum factor

> 'Anyone who is not shocked by quantum theory has not understood it.'
>
> Niels Bohr

The arguments deployed in the previous two chapters suggest that the mind, while not a 'thing' in the usual sense of an entity located at a place with a certain constitution, nevertheless has real existence as an abstract 'high-level' concept in nature's hierarchy of structure. The relation between body and mind, that ancient philosophical enigma, is like the relation between hardware and software in computing. But the connection is tighter than it is in routine computer programming, in the sense that the software (the 'program') is coupled to, or interwoven with, the hardware in what Hofstadter has called a 'Tangled Hierarchy' or 'Strange Loop'. This mosaic of self-reference is the essential feature of consciousness.

The idea of coupling between hardware and software, brain and mind, or matter and information, is not new to science. In the 1920s a revolution occurred in fundamental physics that shook the scientific community and focussed attention as never before on the relation between the observer and the external world. Known as the quantum theory, it forms a pillar in what has become known as the new physics, and provides the most convincing scientific evidence yet that consciousness plays an essential role in the nature of physical reality.

Considering that the quantum theory is now several decades old, it is remarkable that its stunning ideas have taken so long to percolate through to the layman. There is, however, a growing awareness that the theory contains some astonishing insights into the nature of the mind and the reality of the external world, and that full account must be taken of the quantum revolution in the search for an understanding of God and existence. Many modern writers are finding close parallels between the concepts used in the quantum theory and those of Oriental mysticism, such as Zen. But whatever one's religious persuasions, the quantum factor cannot be ignored.

Before embarking on a discussion of these issues it must be made clear that the quantum theory is primarily a practical branch of physics, and as such is brilliantly successful. It has given us the laser, the electron microscope, the transistor, the superconductor and nuclear power. At a stroke it explained chemical bonding, the structure of the atom and nucleus, the conduction of electricity, the mechanical and thermal properties of solids, the stiffness of collapsed stars and a host of other important physical phenomena. The theory has now penetrated most areas of scientific inquiry, at least in the physical sciences, and for two generations has been learned as a matter of course by most science undergraduates. These days, it is applied in many routine practical ways in engineering. In short, the quantum theory is, in its everyday application, a very down-to-Earth subject with a vast body of supporting evidence, not only from commercial gadgetry, but from careful and delicate scientific experiments.

Even though few professional physicists stop to think about the bizarre philosophical implications of the quantum theory, the truly weird nature of the subject emerged very soon after its inception. The theory arose from attempts to describe the behaviour of atoms and their constituents, so it is primarily concerned with the microworld.

Physicists had known for some time that certain processes, such as radioactivity, seem to be random and unpredictable. While large numbers of radioactive atoms obey the laws of statistics, the exact moment of decay of an individual atomic nucleus cannot be predicted. This fundamental uncertainty extends to all atomic and subatomic phenomena, and requires a radical revision of commonsense beliefs to explain it. Before atomic uncertainty was discovered in the early part of this century, it was assumed that all material objects complied strictly with the laws of mechanics, which operate to keep the planets in their orbits, or direct the bullet towards its target. The atom was considered to be like a scaled-down version of the solar system, with its internal components turning like precision clockwork. It turned out to be illusory. In the 1920s it was discovered that the atomic world is full of murkiness and chaos. A particle such as an electron does not appear to follow a meaningful, well-defined trajectory at all. One moment it is found here, the next there. Not only electrons, but all known subatomic particles — even whole atoms — cannot be pinned down to a specific motion. Scrutinized in detail, the concrete matter of daily experience dissolves in a maelstrom of fleeting, ghostly images.

Uncertainty is the fundamental ingredient of the quantum theory. It leads directly to the consequence of *unpredictability*. Does every event have a cause? Few would deny it. In Chapter 3 it was explained how the cause-effect chain has been used to argue for the existence of God — the first cause of everything. The quantum factor, however, apparently breaks the chain by allowing effects to occur that have no cause.

Already in the twenties, controversy raged over the meaning behind the unpredictable face of atoms. Is nature inherently capricious, allowing electrons and other particles to simply pop about at random, without rhyme or reason — events without a cause? Or are these particles like corks being tossed about by an unseen ocean of microscopic forces?

Most scientists, under the leadership of the Danish physicist Niels Bohr, accepted that atomic uncertainty is truly intrinsic to nature: the rules of clockwork might apply to familiar objects such as snooker balls, but when it comes to atoms, the rules are those of roulette. A dissenting, albeit distinguished, voice was that of Albert Einstein. 'God does not play dice', he declared. Many ordinary systems, such as the stock market or the weather, are also unpredictable. But that is only because of our ignorance. If we had complete knowledge of all the forces concerned, we could (in principle at least) anticipate every twist and turn.

The Bohr–Einstein debate is not just one of detail. It concerns the entire conceptual structure of science's most successful theory. At the heart of the subject lies the bald question: is an atom a *thing*, or just an abstract construct of imagination useful for explaining a wide range of observations? If an atom *really* exists as an independent entity then at the very least it should have a location and a definite motion. But the quantum theory denies this. It says that you can have one or the other but not both.

This is the celebrated uncertainty principle of Heisenberg, one of the founders of the theory. It says you can't know where an atom, or electron, or whatever, is located *and* know how it is moving, at one and the same time. Not only can you not know it, but the very concept of an atom with a definite location and motion is meaningless. You can ask where an atom is and get a sensible answer. Or you can ask how it is moving and get a sensible answer. But there is no answer to a question of the sort 'Where is it and how fast is it going?' Position and motion (strictly, momentum) form two mutually incompatible aspects of reality for the microscopic particle. But what right have we to say that an atom is a *thing* if it isn't located somewhere, or else has no meaningful motion?

According to Bohr, the fuzzy and nebulous world of the atom only sharpens into concrete reality when an observation is made. In the absence of an observation, the atom is a ghost. It only materializes when you look for it. And you can decide what to look for. Look for its location and you get an atom at a place. Look for its motion and you get an atom with a speed. But you can't have both. The reality that the observation sharpens into focus cannot be separated from the observer and his choice of measurement strategy.

If all this seems too mind-boggling or paradoxical to accept, Einstein would have agreed with you. Surely the world out there really exists whether we observe it or not? Surely everything that happens does so for its own reasons and not because it is being watched? Our observations might uncover the atomic reality, but how can they *create* it? True, atoms and their components might seem to behave in a muddled and imprecise way, but that is only due to our clumsiness in probing such delicate objects.

The essential dichotomy can be illustrated with the aid of the humble television. The image on a television screen is produced by myriads of light pulses emitted when electrons fired from a gun at the back of the set strike the fluorescent screen. The picture you perceive is reasonably sharp because the number of electrons involved is enormous, and by the law of averages, the cumulative effect of many electrons is predictable. However, any particular electron, with its inbuilt unpredictability, could go anywhere on the screen. The arrival of this electron at a place, and the fragment of picture that it produces, is uncertain. According to Bohr's philosophy bullets from an ordinary gun follow a precise path to their target, but electrons from an electron gun simply turn up at the target. And however good your aim, no bull's-eye is guaranteed. The event 'electron at place x on the television screen' cannot be considered as *caused* by the gun, or anything else. For there is no known reason why the electron should go to point x rather than some other place. The picture fragment is an event without a cause, an astonishing claim to remember when you next watch your favourite programme.

Nobody is saying, of course, that the electron gun has nothing whatever to do with the electron's arrival, only that it does not completely determine it. Instead of envisaging the electron at the target as really existing prior to its arrival and being connected to the gun by a precise trajectory, physicists think of the electron leaving the gun as being in a sort of limbo, its presence represented only by cohorts of ghosts. Each ghost explores its own path to the screen, though only one electron actually appears on the screen itself.

How can these weird ideas be confirmed?

11 The decay of an atom, or a subnuclear particle, can produce two oppositely spinning particles (e.g. photons of light) which travel in opposite directions, perhaps to a large distance.

In the 1930s Einstein conceived of an experiment which he believed would expose the fraud of the quantum ghosts, and establish once and for all that every event has a distinct cause. The experiment is based on the principle that the multitude of ghosts do not act independently, but in collusion. Suppose, said Einstein, that a particle explodes into two fragments, and these fragments are allowed to travel undisturbed, a long way apart. Although well separated, each fragment will carry an imprint of its partner. For example, if one flies off spinning clockwise the other, by reaction, will spin anti-clockwise.

The ghost theory claims that each fragment will be represented by more than one potential possibility. To pursue the example, fragment A will have two ghosts, one spinning clockwise, the other anti-clockwise. Which ghost becomes the *real* particle has to await a definite measurement or observation. Similarly, the oppositely-moving partner, fragement B, will also be represented by two counter-spinning ghosts. However, if a measurement of A promotes, for instance, the clockwise ghost to reality, B has no choice: it must promote its anti-clockwise ghost. The two separated ghost particles must co-operate with each other to comply with the law of action and reaction (see Fig. 11).

It seems baffling, to say the least, how fragment B can possibly *know* for which of its two ghosts fragment A has opted. If the fragments are well separated, it is hard to see how they can communicate. Further-

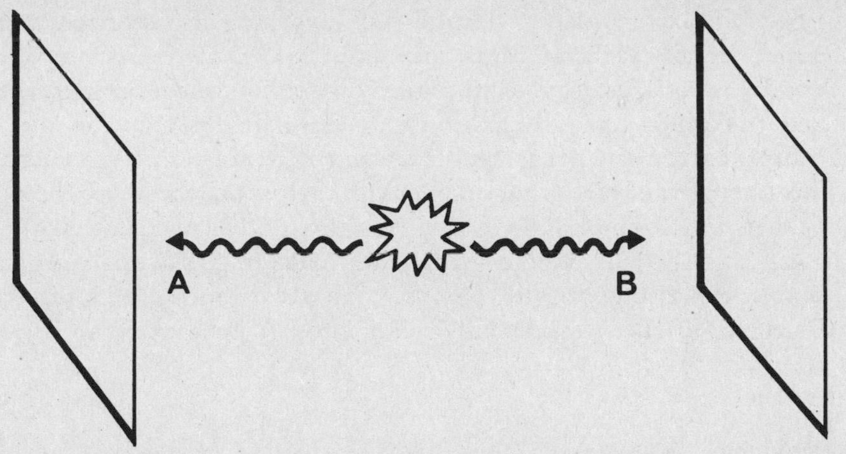

12 If two photons with correlated spins and polarization encounter parallel pieces of polarizing material, they will show 100 per cent cooperation: whenever photon A is blocked, so is B. This cooperation occurs even though (i) the actual outcome of the encounter between photon and polarizer is completely unpredictable, and (ii) the photons may be very far apart.

more, if both fragments are observed simultaneously, there is simply no time for any signal to propagate between the fragments. Einstein insisted that this result is paradoxical unless the fragments *really* exist (already spinning in a particular way) at the instant they separate, and that they retain their spin during their flight apart. There are no ghosts; there is no delay of choice until measurement, no mysterious co-operation without communication.

Bohr replied that Einstein's reasoning assumed the two fragments are independently real because they are well separated. In fact, asserted Bohr, it is not possible to regard the world as made up of lots of separated bits. Until a measurement is actually performed, both A and B must be regarded as a single totality, even if they are light years apart. This is holism indeed!

The real test of Einstein's challenge had to await post-war developments. In the 1960s, the physicist John Bell proved a most remarkable theorem about experiments of the Einstein type. He showed, quite generally, that the degree of cooperation between separated systems cannot exceed a certain definite maximum, if along with Einstein, one assumes that the fragments really do exist in well-defined states prior to their observation. In contrast, the quantum theory predicts that this limit can be exceeded. What was needed was an experiment.

Advances in technology enabled experimental tests to be conducted to check Bell's inequality. Several such experiments have been performed, but by far the best was carried out in 1982 at the University of Paris by Alaine Aspect and colleagues. For subatomic fragments they used two photons of light emitted simultaneously by an atom. Stationed in the path of each photon was a piece of polarizing material. This filters out photons that do not align their vibrations with the axis of the material. Thus, only ghost photons with the correct orientation (polarization) will emerge from the polarizing material. Again, photons A and B cooperate, because their polarizations are forced by action and reaction to be parallel. If photon A is blocked, so is B.

13 Testing Bell's inequality: if the polarizers are oriented obliquely, the cooperation between A and B declines — sometimes A is passed when B is blocked. However, there is found to be some residual cooperation; more than can be explained by any theory that assumes (i) the independent reality of the external world, and (ii) no secret reversed-time communication between the widely-separated photons.

The real test comes when the two pieces of polarizing material are oriented obliquely to each other. The cooperation then declines because the polarizations of the photons cannot now both be aligned with their respective polarizers. And it is here that the Bohr–Einstein controversy can be settled. Einstein's theory predicts considerably less cooperation than Bohr's.

So what was the result?

Bohr wins, Einstein loses. The Paris experiment, taken together with other less accurate experiments performed during the seventies, leaves little room for doubt that the uncertainty of the microworld is intrinsic. Events without causes, ghost images, reality triggered only by observation — all must apparently be accepted on the experimental evidence.

What are the implications of this stunning conclusion?

So long as nature's rebelliousness is restricted to the microworld, many people will feel only slightly uneasy that the concrete reality of the world 'out there' has dissolved. In daily life a chair is still a chair, is it not?

Well — not quite.

Chairs are made of atoms. How can lots of ghosts combine together to make something real and solid? And what about the observer himself? What is so special about a human being that gives him the power to focus the fuzziness of atoms into sharp reality? Does an observer have to be human? Will a cat suffice, or a computer?

The quantum theory is one of the most difficult and technical subjects to understand, and this brief review can do no more than lift a small corner of the veil of mystery to allow the reader a glimpse of its bizarre concepts. (The subject is treated in much greater detail in my book *Other Worlds.*) This sketchy survey will, however, demonstrate that the commonsense view of the world, in terms of objects that really exist 'out there' independently of our observations, totally collapses in the face of the quantum factor.

Many of the perplexing features of the quantum theory can be understood in terms of a curious 'wave-particle' duality, reminiscent of the mind–body duality. According to this idea, a microscopic entity such as an electron or a photon sometimes behaves like a particle and sometimes like a wave; it depends on the sort of experiment chosen. A particle is a totally different animal from a wave: it is a small lump of concentrated stuff, whereas a wave is an amorphous disturbance that can spread out and dissipate. How can anything be both?

It all has to do with complementarity again. How can the mind be both thoughts and neural impulses? How can a novel be both a story and a collection of words? Wave-particle duality is another software–hardware dichotomy. The particle aspect is the hardware face of atoms — little balls rattling about. The wave aspect corresponds to software, or mind, or information, for the quantum wave is not like any other sort of wave anybody has ever encountered. It is not a wave of any substance or physical stuff, but a wave of knowledge or information. It is a wave that tells us what can be known about the atom, not a wave of the atom itself. Nobody is suggesting that an atom can ever spread itself around as an undulation. But what can spread itself around is what an observer can know about the atom. We are all familiar with crime waves; not waves of any substance but waves of *probability*. Where the crime wave is most intense, there is the greatest likelihood of a felony.

The quantum wave is also a wave of probability. It tells you where you can expect the particle to be, and what chance it may have of such-and-such a property, such as rotation or energy. The wave thus encapsulates the inherent uncertainty and unpredictability of the quantum factor.

No experiment better illustrates the conflict and dichotomy of wave-particle duality than Thomas Young's two-slit system. Light, according to the long tradition of classical physics, is a wave — an electromagnetic wave, an undulation of the electromagnetic field. About 1900, however, Max Planck demonstrated mathematically that light waves can behave in some ways like particles — we now call them photons. Light, according to Planck, comes in indivisible lumps or packets (hence the Latin word quantum). The idea was refined by Einstein, who pointed out that these corpuscular photons can knock electrons out of atoms after the fashion of a coconut shy. This is what happens in the now familiar photocell; odd, but not outrageous.

The first unexpected twist comes when two light beams are combined together. If two wave systems are superimposed, an effect called interference results. Imagine two stones dropped a few inches apart into a still pond. Where the spreading disturbances overlap a complex pattern of undulations occurs. In some regions the two wave motions come together in phase and the disturbance is amplified; elsewhere the waves meet out of phase and cancel each other.

To get the same effect with light we can illuminate two holes in a screen. The light waves spilling through each aperture spread out and overlap, creating an interference pattern which is readily revealed by a photographic plate. The image of the two holes is not merely two fuzzy blobs, but a systematic pattern of bright and dark patches, indicating where the two wave trains have arrived in step, and out of step, respectively (see Fig. 14).

All this was well known in the early nineteenth century. Strange overtones develop, however, when the corpuscular nature of light is taken into account. Each photon hits the photographic plate at a particular point and makes a little spot. The extended image, as in the television case, therefore builds up from millions of speckles as the photons strike the plate like a hail of shot. The point of arrival of any individual photon is definitely unpredictable. All we know is that there is a good chance it will hit the plate in a bright-patch area.

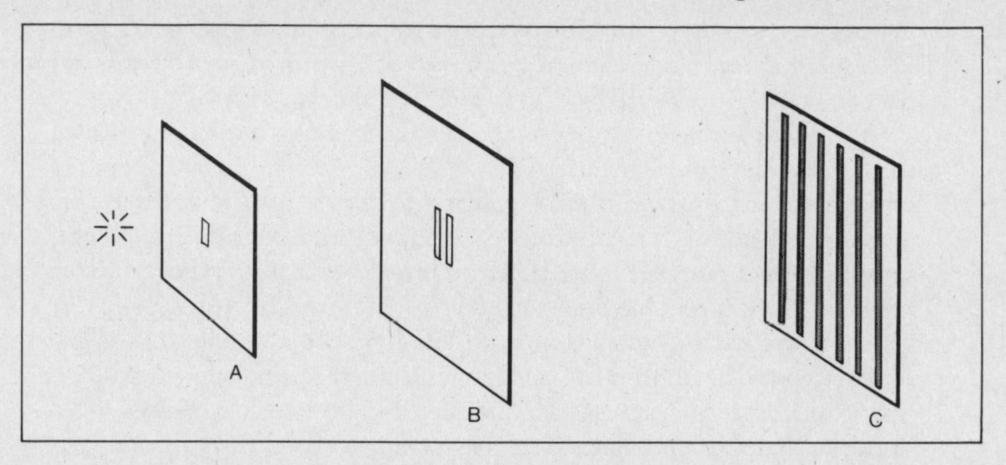

14 The famous Young's two-slit experiment is ideal for exposing the bizarre wave-particle duality of light (it can also be performed with electrons or other particles). The small hole in screen A illuminates the two narrow slits in screen B. The image of the slits is displayed on screen C. Rather than a simple double band of light, there appears a sequence of bright and dark bands (interference fringes) caused by the light waves from each slit arriving successively in step or out of step, depending on position. Even when one photon at a time traverses the apparatus, the same interference pattern builds up in a speckled fashion, though any given photon can only go through *either* one slit *or* the other in screen B, and has no neighbouring photons against which to gauge its 'step'.

That, however, is not all. Suppose we turn down the illumination so that only *one photon at a time* passes through the system. Given long enough, the accumulated speckles will still build up to give the bright and dark interference pattern. The paradox is that any particular photon can surely pass through only *one* of the holes? Yet the interference pattern requires *two* overlapping wave trains, one from each hole. The entire experiment can, in fact, be performed with atoms, electrons or other subatomic particles instead of light. In all cases an interference pattern composed of individual speckles results, demonstrating that photons, atoms, electrons, mesons, and so on manifest both wave and particle aspects.

In the 1920s Bohr gave a possible resolution of the paradox. Think of the case when the photon goes through hole A as one possible world (world A) and the route through hole B as another (world B). Then *both* these worlds, A and B, are somehow present together, superimposed. We cannot say, Bohr asserted, that the world of our experience represents *either* A *or* B, but is a genuine hybrid of the two.

Moreover, this hybrid reality is not simply the sum total of the two alternatives, but a subtle marriage: each world *interferes* with the other to produce the celebrated pattern. The two alternative worlds overlap and combine, rather like two movie films being projected simultaneously onto the same screen.

Einstein, the eternal sceptic, refused to accept hybrid realities. He confronted Bohr with a modified version of the two-hole experiment, in which the screen is allowed to move freely. Careful observation, he insisted, would enable one to determine through which hole the photon went. Passage through the left-hand hole results in a slight deflection of the photon to the right, and the recoiling screen could in principle be seen to move to the left. Motion to the right would indicate that the other hole had been traversed. By this means, experiment would determine that *either* world A *or* B corresponds to reality. Furthermore, the apparent indeterminacy of the photon's behaviour in the original experiment could then simply be attributed to the coarseness of the experimental technique in that arrangement.

Bohr countered decisively. Einstein was changing the rules in midgame. If the screen is free to move, then its motion is also subject to the inherent uncertainty of quantum physics. Bohr easily showed that the effect of recoil would be to destroy the interference pattern on the photographic plate, producing merely two fuzzy blobs instead. Either the screen is clamped, and the wavelike nature of light is manifested in the interference pattern, or the screen is freed, and a definite trajectory for the photon is established. But in that case the wave-like aspect disappears, and the light behaves in a purely corpuscular way. We are thus dealing with two different experiments. They are not contradictory, but complementary. Einstein's strategy tells us nothing about the photon paths in the original experiment, where the hybrid world is manifested.

The bizarre conclusion from this exchange is that we — the experimenters — are involved in the nature of reality in a fundamental way. By choosing to clamp the screen we can construct a mysterious hybrid world in which photon paths have no well-defined meaning.

In 1979, John Wheeler, speaking ironically at a symposium in Princeton celebrating Einstein's centenary, drew a still more mind-boggling conclusion from the two-hole experiment. He pointed out that by a simple modification of the apparatus it is possible to delay the choice of measurement strategy until *after* the photon has passed through the screen. Our decision to make a hybrid world can thus be delayed until after that world has come into existence! The precise nature of reality, Wheeler claims, has to await the participation of a conscious observer. In this way, mind can be made responsible for the retroactive creation of reality — even a reality that existed before there were people. This is the retroactive causation mentioned on page 39.

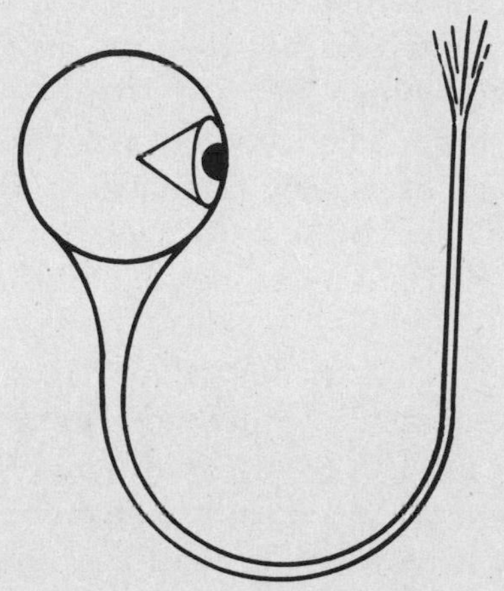

15 This symbolic picture due to John Wheeler represents the universe as a self-observing system. Wheeler's astonishing modification of the Young's two-slit experiment reveals that an observer today can be made partially responsible for generating the reality of the remote past. The tail of the figure can thus represent the early stages of the universe, being promoted to concrete reality through its later observation by consciousness which itself depends on that reality.

It will be evident from the foregoing that the quantum theory demolishes some cherished commonsense concepts about the nature of reality. By blurring the distinction between subject and object, cause and effect, it introduces a strong holistic element into our world view. We have seen how, in the Einstein experiment, two widely separated particles must nevertheless be regarded as a single system. We have also seen how it is meaningless to talk about the condition of an atom, or even the very notion of an atom, except within the context of a specified experimental arrangement. To ask where an atom is *and* how it moves is forbidden. First establish what you want to measure — position or motion — then you will get a sensible answer. The measurement will involve large chunks of macroscopic apparatus. Thus the microscopic reality is inseparable from the macroscopic reality. Yet the macroscopic is made up of the microscopic — apparata are made of atoms! Strange loops again.

David Bohm, a leading quantum theorist, addressed these issues in his book *Wholeness and the Implicate Order*:

> A centrally relevant change in descriptive order required in the quantum theory is thus the dropping of the notion of analysis of the world into relatively autonomous parts, separately existent but in interaction. Rather, the primary emphasis is now on *undivided wholeness*, in which the observing instrument is not separated from what is observed.[1]

In short, the world is not a collection of separate but coupled *things*; rather it is a network of *relations*. Bohm here echoes the words of Werner Heisenberg: 'The common division of the world into subject and object, inner world and outer world, body and soul is no longer adequate.'

How can we resolve the paradoxical loop that the macroworld — the world of daily experience — determines the microscopic reality that it is, itself, made of? This issue is confronted head on when we ask what actually happens when a quantum measurement is made. How does the observer contrive to project the fuzzy microworld into a state of concrete reality?

The quantum 'measurement problem' is really a variant of the mind-body or software–hardware problem; physicists and philosophers have struggled with it for decades. The hardware — the particle — is described by a wave, which encodes the information (software) about what an observer is likely to find the particle doing when he observes it. When an observation is made the wave 'collapses' into a particular state that ascribes a definite sharp value to whatever has been observed.

Paradoxes arise when the act of measurement is described throughout at purely the hardware level. Suppose an electron scatters off a target. It could go either right or left. You compute with the wave and find out where the wave goes. The wave diffracts off the target and spreads out, partly to the right and partly to the left, with equal strength, for instance. This means a fifty–fifty chance that, on observation, you will find the electron on *either* the left *or* the right. It is important to remember, though, that until the observation is actually performed, it is not possible to say (or indeed to meaningfully discuss) on which side of the target the electron *really* is located. The electron keeps its options open until you actually peek. Both possible worlds coexist in a hybrid, ghostly superposition (see Fig. 16).

Now you make your observation, and the electron is found, say, on the left. Instantly the right hand 'ghost' vanishes. The wave suddenly

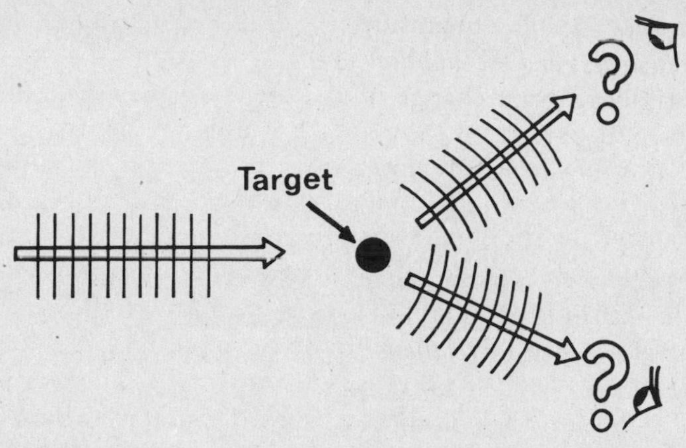

16 Described as a wave, an electron ricochets off the target by producing ripples that travel both leftwards and rightwards. Until an observation is made of where the electron has deflected, it is necessary to suppose that two ghost-worlds (or ghost-electrons) coexist in a hybrid state of unreality. At the instant of observation one of the ghosts disappears, its associated wave simply collapsing, and the electron is promoted from its former state of limbo to a single concrete reality. Mystery surrounds just what it is that the observer does to the electron to achieve this abrupt promotion. Is it mind-over-matter? Does the universe split into two parallel realities?

collapses over to the left hand side of the target, for there is now no possibility of the electron being on the right. What causes this dramatic collapse?

In order to make an observation it is necessary to couple the electron to a piece of external apparatus, or perhaps a series of apparata. These have the job of sniffing out where the electron is and amplifying the signal up to the macroscopic level where it can be recorded. But these couplings and apparatus processes are themselves mechanical activities involving atoms (albeit in large numbers), and are therefore subject to the quantum factor too. We could write down a wave to represent the measuring apparatus. Suppose that the measuring machine is equipped with a pointer which has two positions, one to indicate that the electron is on the left, the other implying it is on the right. Then viewing the total system of electron plus apparatus as a large quantum system forces us to conclude that the hybrid nature of the balking electron is transferred to the pointer. Instead of the measurement device showing either one pointer position or the other, it ought to go into a state of quantum limbo. In this way, a measurement seems to amplify the nightmare quantum world up to laboratory scale.

This paradox was investigated by the mathematician John von Neumann, who demonstrated (using a simple mathematical model) that the effect of coupling the electron to the measuring apparatus does indeed prod the electron into opting for either left or right, but at the price of transferring the hybrid unreality to the apparatus pointer. Von Neumann also showed, however, that if the apparatus is in turn coupled to another instrument that reads the output of the first instrument, then the first pointer would thereby be prodded into a decision too. But now the second apparatus goes into limbo. There can thus be a whole chain of machines looking at each other and recording sensible 'either–or' results, but always the last member of von Neumann's chain will be left in a state of unreality.

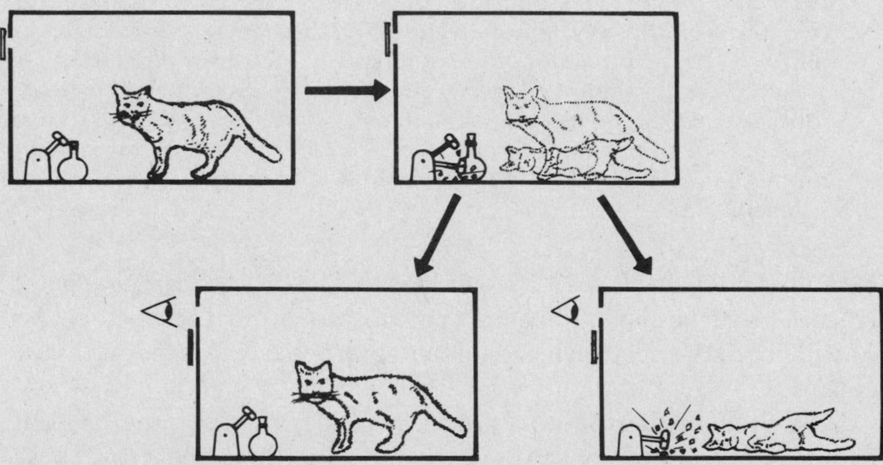

17 The sad tale of Schrödinger's cat. A quantum process can trigger the release of cyanide with a 50:50 probability. Quantum theory requires that the system develops into a ghost-like hybrid state of live–dead cat until an observation is made, when *either* a live cat *or* a dead cat will be perceived. This thought experiment highlights the weird implications surrounding the act of observation in the quantum theory.

The eccentric consequences are highlighted by a famous paradox due to Schrödinger in which the amplifying device is used to trigger the release of a poison which can kill a cat. The left–right pointer dichotomy thus becomes a live–dead cat dichotomy. If a cat is to be described as a quantum system one is forced to conclude that, until the cat is observed by someone or something, it is suspended in a schizophrenic 'live–dead' condition, which seems absurd.

Suppose we use a person instead of a cat. Do they experience a live–dead state? Of course not. So has quantum mechanics broken down when it comes to human observers? Does von Neumann's chain end when it reaches the consciousness of a person? This sensational claim has indeed been made by a leading quantum theorist, Eugene Wigner. Wigner suggests that it is the entry of the information about the quantum system into the mind of the observer that collapses the quantum wave and abruptly converts a schizophrenic, hybrid, ghost state into a sharp and definite state of concrete reality. Thus, when the experimenter himself looks at the apparatus pointer, he causes it to decide upon either one position or the other, and thereby, down the chain, also forces the electron to make up its mind.

If Wigner's thesis is accepted it returns us to the old idea of dualism — that mind exists as a separate entity on the same level as matter and can act on matter causing it to move in apparent violation of the laws of physics. Wigner serenely accepts this: 'Does the consciousness influence the physico-chemical conditions (of the brain)? In other words, does the human body deviate from the laws of physics, as gleaned from the study of inanimate matter? The traditional answer to this question is, "No": the body influences the mind but the mind does not influence the body. Yet at least two reasons can be given to support the opposite thesis.'[2] One of these two reasons Wigner cites is the law of action and reaction. If body acts on mind the reverse should also be true. The other is the aforementioned resolution of the quantum measurement problem that results.

It has to be admitted that very few physicists support Wigner's ideas, though some have seized upon the quantum route to mind-over-matter to argue for the acceptability of certain paranormal phenomena, such as psychokinesis and remote metal-bending. ('If the mind can fire neurons why can't it bend spoons?')

There is a strong hint of level-confusion running through the Wigner thesis. The attempt to discuss the operation of hardware (electrons running about) by appeal to software (the mind) falls into the dualist trap. However, the issue is more subtle here because hardware and software are hopelessly entangled in the quantum theory (for example, in wave-particle duality). Whatever the validity of Wigner's ideas, they do suggest that the solution of the mind–body problem may be closely connected with the solution of the quantum measurement problem, whatever that will eventually be.

Another attempt to break out of the quantum measurement paradox is perhaps even more bizarre than Wigner's appeal to the mind. So long as one is dealing with a finite physical system, von Neumann's chain can be extended. You can always claim that everything you perceive is real because there exists a larger system which collapses what you see into reality by 'measuring' or 'observing' it. But in recent years physicists have been interested in the subject of quantum cosmology — the quantum theory of the entire universe. By definition, there can be nothing outside the universe to collapse the whole cosmic panorama into concrete existence (except God, perhaps?). At this level, the universe would seem to be caught in a state of limbo or cosmic schizophrenia. Without a Wigner-type mind to integrate it, the universe seems destined to languish as a mere collection of ghosts, a multi-hybrid superposition of overlapping alternative realities, none of them the *actual* reality. Why, then, do we perceive a single, concrete reality?

One bold idea addresses this unnerving issue face on: the parallel universe theory. Invented by physicist Hugh Everett in 1957, and subsequently championed by Bryce DeWitt, now at the University of Texas at Austin, the theory proposes that all the possible alternative quantum worlds are equally real, and exist in parallel with one another. Whenever a measurement is performed to determine, for example, whether the cat is alive or dead, the universe divides into two, one containing a live cat, the other a dead one. Both worlds are equally real, and both contain human observers. Each set of inhabitants, however, perceives only their own branch of the universe.

Commonsense may rebel against the extraordinary concept of the universe branching into two as the result of the antics of a single electron, but the theory stands up well to closer scrutiny. When the universe splits, our minds split with it, one copy going off to populate each world. Each copy thinks it is unique. Those who object that they don't feel themselves being split should reflect on the fact that they do not feel the motion of the Earth around the sun either. The splitting is repeated again and again as every atom, and all the subatomic particles, cavort about. Countless times each second, the universe is replicated. Nor is it necessary for an actual measurement to be performed in order that the replication occur. It suffices that a single microscopic particle merely interacts in some way with a macroscopic system. In the words of DeWitt:

> Every quantum transition taking place on every star, in every galaxy, in every remote corner of the universe is splitting our local world on earth into myriads of copies of itself . . . Here is schizophrenia with a vengeance.[3]

The price paid for the restoration of reality is a multiplicity of realities — a stupendous and growing number of parallel universes, diverging along their separate branches of evolution.

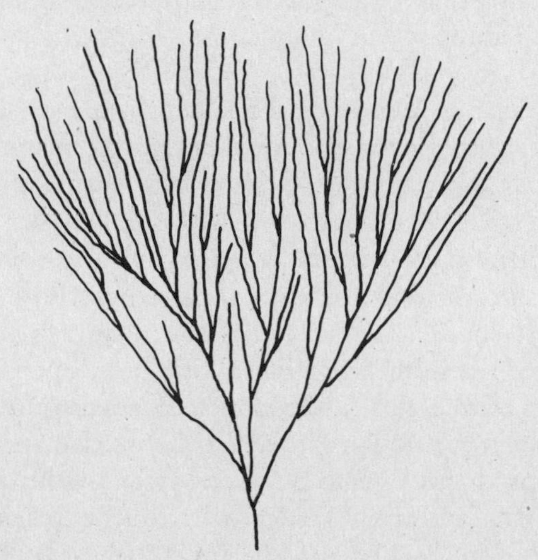

18 To avoid live–dead cats and other quantum schizophrenic unrealities, Everett proposed that the indeterminism of quantum systems generates a multifoliate reality in which the universe is continually branching into myriads of 'parallel universes', physically disconnected, but equally real. The observer's mind is also split into countless duplicates by this process.

What are these other worlds like? Can we travel to them? Do they explain UFOs or the mysterious disappearances in the Bermuda triangle? Alas for the ufologists, the Everett theory is explicit on this point. The parallel worlds, once disconnected, are physically isolated for all practical purposes. To reunite them would require reversing a measurement, which amounts to reversing time. It would be rather like reconstituting a broken egg, atom by atom.

But where *are* these worlds? In a sense, those that closely resemble our own are very nearby. Yet they are totally inaccessible: we cannot reach them however far we travel through our own space and time. The reader of this book is no more than an inch away from millions of his duplicates, but that inch is not measured through the space of our perceptions.

The farther apart the worlds have branched, the greater their differences. Worlds that split away from our own in some trivial way, such as the path of a photon in a two-hole experiment, would be indistinguishable to the casual glance. Others would differ in their cat populations. In some worlds Hitler would not have been, John Kennedy lives on. Yet others would be wildly different, especially those that branched away from each other near the beginning of time. In fact, everything that could possibly happen (though not everything that can conceivably happen) does happen somewhere, in some branch of this multifoliate reality.

The simultaneous existence of all possible worlds raises the intriguing question of why the world in which this book is being read is the one it is, and not one of the other, very different branches. Obviously, the reader cannot exist in all or indeed the vast majority of the other worlds, for their conditions are simply not suitable for life to arise. (We shall return to this issue in Chapter 12.)

Many people have suggested that the quantum theory, with its involvement of the mind in so basic a fashion, opens the door to an understanding of free will. The old idea of a deterministic universe in which everything we do is decided by the mechanics of the universe long before our birth seems to be swept away by the quantum factor. So is free will alive and well? To deal with this matter properly we first have to delve more deeply into the mysteries of time.

9. Time

'There is not even a meaning to the word *experience* which would not presuppose the distinction between past and future.'

Carl von Weizsäcker

'But at my back I always hear
Time's wingèd chariot hurrying near'

Andrew Marvell

Two great revolutions gave birth to the new physics: the quantum theory and the theory of relativity. The latter, almost exclusively due to the work of Einstein, is a theory of space, time and motion. Its consequences are as equally baffling and profound as the quantum theory, and challenge many cherished notions about the nature of the universe. Never is this more so than in the theory's treatment of time — a subject of intense and longstanding concern in all the world's great religions.

Time is so fundamental to our experience of the world that any attempt to tinker with it meets with great scepticism and resistance. Every week I receive manuscipts by amateur scientists intent on finding fault with Einstein's work, attempting to restore the common-sense, traditional concept of time despite almost eighty years of success during which not a single experiment has marred the flawless predictions of the theory of relativity.

Our very notion of personal identity — the self, the soul — is closely bound up with memory and *enduring* experience. It is not sufficient to proclaim 'I exist', at this instant. To *be* an individual implies a continuity of experience together with some linking feature, such as memory. The strong emotional and religious overtones of the subject probably account both for the resistance to the claims of the new physics and for the deep fascination which scientists and laymen alike share for the mind-bending consequences of the theory of relativity.

The so-called special theory of relativity, published in 1905, arose from attempts to reconcile an apparent conflict between the motion of material bodies and the propagation of electromagnetic disturbances. In particular, the behaviour of light signals seem to be in flagrant violation of the long-standing principle that all uniform motion is purely relative. The technical details need not concern us here. The result was that Einstein restored the relativity principle, even for the case when light signals are involved, but at a price.

The first casualty of the special theory was the belief that time is absolute and universal. Einstein demonstrated that time is, in fact, elastic and can be stretched and shrunk by motion. Each observer carries around his own personal scale of time, and it does not generally agree with anybody else's. In our own frame, time never appears distorted, but relative to another observer who is moving differently, our time can be wrenched out of step with their time.

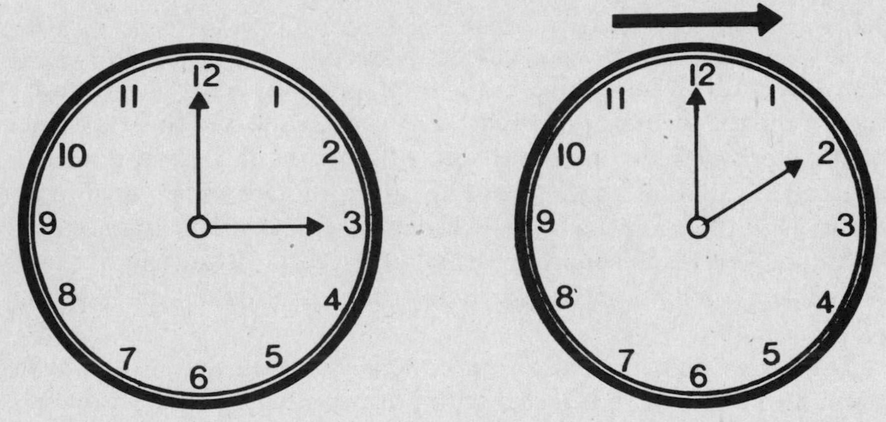

19 The time dilation effect, now a routine experience for physicists, can be demonstrated by using rapidly moving, sensitive atomic clocks, or subatomic particles with known decay rates. The moving clock runs slow relative to its neighbour. This leads to the famous 'twins effect' in which an astronaut returns from a high-speed voyage some years younger than his Earthbound twin.

This weird dislocation of time scales opens the way to a type of time travel. In a sense, we are all travellers in time, heading towards the future, but the elasticity of time enables some people to get there faster than others. Rapid motion enables you to put the brakes on your own time scale, and let the world rush by, as it were. By this strategy it is possible to reach a distant moment more quickly than by sitting still. In principle one could reach the year 2000 in a few hours. However, to achieve an appreciable timewarp speeds of many thousands of miles per second are necessary. At currently available rocket speeds only precision atomic clocks can reveal the minute dilations. The key to these effects is the speed of light. As it is approached, so the timewarp escalates. The theory forbids anyone to break the light barrier, which would have the effect of turning time inside out.

It is possible to telescope time dramatically using high-speed subatomic particles. Whirled about in a giant accelerator very close to the speed of light, particles called muons have been 'kept alive' for dozens of times longer than would be expected if they were at rest (when they decay in about a microsecond).

Equally extraordinary effects afflict space, which is also elastic. When time is stretched, space is shrunk. Rushing on a train through a railway station, the station clock runs slightly slower as viewed from your frame of reference, relative to that of a porter on the platform. In compensation, the platform appears to you to be somewhat shorter. Of course we never notice such effects because they are too small at ordinary speeds, but they are easily measured on sensitive instruments. The mutual distortions of space and time can be regarded as a conversion of space (which shrinks) into time (which stretches). A second of time, however, is worth an awful lot of space — about 186,000 miles to be precise.

Time distortions of this sort are a favourite sci-fi gimmick, but there is nothing fictional about them. They really do occur. One bizarre phenomenon is the so-called twins effect. An itinerant twin blasts off to a nearby star, nudging the light barrier. The stay-at-home twin waits for him to return ten years later. When the rocket gets back, the Earth-bound twin finds his brother has aged only one year to his ten. High speed has enabled him to experience only one year of time, during which ten years have elapsed on Earth.

Einstein went on to generalize his theory to include the effects of gravity. The resulting general theory of relativity incorporates gravity, not as a force, but a distortion of spacetime geometry. In this theory, spacetime is not 'flat', obeying the usual rules of school geometry, but curved or warped, giving rise to both spacewarps and timewarps.

As discussed in Chapter 2, modern instruments are so sensitive that even the Earth's gravitational timewarp can be detected by clocks in rockets. Time really does run faster in space, where the Earth's gravity is weaker.

20 Gravity slows time, as may be demonstrated experimentally even on Earth. The clock at the top of the tower gains relative to that at the base.

The stronger the gravity, the more pronounced is the timewarp. Some stars are known where the grip of gravity is so ferocious that time there is slowed by several per cent relative to us. In fact, these stars are on the brink of the threshold at which runaway timewarps set in. If the gravity of such a star were a few times greater, the timewarp would escalate until, at a critical value of the gravity, time would grind to a halt altogether. Viewed from Earth, the surface of the star would be frozen into immobility. We could not, however, see this extraordinary temporal suspension because the light by which we would view it is also seized by the same torpidity, and its frequency depressed beyond the visible region of the spectrum. The star would appear black.

Theory suggests that a star in this condition could not remain inert, but would succumb to its own intense gravity and implode in a microsecond to a spacetime singularity, leaving behind a hole in space — a black hole. The timewarp of the erstwhile star remains imprinted in the empty space.

A black hole, therefore, represents a rapid route to eternity. In this extreme case, not only would a rocket-bound twin reach the future quicker, he could reach the *end of time* in the twinkling of an eye! At the instant he enters the hole, all of eternity will have passed outside according to his relative determination of 'now'. Once inside the hole, therefore, he will be imprisoned in a timewarp, unable to return to the outside universe again, because the outside universe will have happened. He will be, literally, beyond the end of time as far as the rest of the universe is concerned. To emerge from the hole, he would have to come out before he went in. This is absurd and shows there is no escape. The inexorable grip of the hole's gravity drags the hapless astronaut towards the singularity where, a microsecond later, he reaches the edge of time, and obliteration; the singularity marks the end of a one-way journey to 'nowhere' and 'nowhen'. It is a nonplace where the physical universe ceases.

The revolution in our conception of time which has accompanied the theory of relativity is best summarized by saying that, previously, time was regarded as absolute, fixed, and universal — independent of material bodies or observers. Today time is seen to be *dynamical*. It can stretch and shrink, warp and even stop altogether at a singularity. Clock rates are not absolute, but relative to the state of motion or gravitational situation of the observer.

Liberating time from the strait-jacket of universality, and allowing each observer's time to roll forward freely and independently, forces us to abandon some long-standing assumptions. For example, there can be no unanimous agreement about the choice of 'now'. In the twins experiment, the rocket twin, during his outward trip, might wonder: 'What is my twin on Earth doing *now*?' But the dislocation of their relative time scales means that 'now' in the frame of the rocket is quite a different moment from 'now' as judged on Earth. There is no universal 'present moment'. If two events, A and B, occurring at separated places, are regarded as simultaneous by one observer, another observer will see A occur before B, while yet another may regard B as occurring first.

The idea that the time order of two events might appear different to different observers seems paradoxical. Can the target shatter before the gun fires? Fortunately for causality, this does not happen. For events A and B to have an uncertain sequence, they must occur within a short enough duration that it would be impossible for light to travel from place A to place B in that interval. In the theory of relativity, light signals make all the rules, and in particular they forbid any influence or signal to travel faster than they do. If light isn't fast enough to connect A and B, nothing is, so A and B cannot influence one another in any way. There is no causal connection between them; reversing the time order of A and B does not amount to reversing cause and effect.

One inevitable victim of the fact that there is no universal present moment is the tidy division of time into past, present and future. These terms may have meaning in one's immediate locality, but they can't apply everywhere. Questions such as 'What is happening *now* on Mars?' are intended to refer to a particular instant on that planet. But as we have seen, a space traveller sweeping past Earth in a rocket who asked the same question at the same instant would be referring to a different moment on Mars. In fact, the range of possible 'nows' on Mars available to an observer near Earth (depending on his motion) actually spans several minutes. When the distance to the subject is greater, so is this range of 'nows'. For a distant quasar 'now' could refer to any interval over billions of years. Even the effect of strolling around on foot alters the 'present moment' on a quasar by thousands of years!

The abandonment of a distinct past, present and future is a profound step, for the temptation to assume that only the present 'really exists' is great. It is usually presumed, without thinking, that the future is as yet unformed and perhaps undetermined; the past has gone, remembered but relinquished. Past and future, one wishes to believe, do not exist. Only one instant of reality seems to occur 'at a time'. The theory of relativity makes nonsense of such notions. Past, present and future must be equally real, for one person's past is another's present and another's future.

The physicist's attitude to time is strongly conditioned by his experiences with the effects of relativity and can appear quite alien to the layman, although the physicist himself rarely thinks twice about it. He does not regard time as a sequence of events which *happen*. Instead, all of past and future are simply *there*, and time extends in either direction from any given moment in much the same way as space stretches away from any particular place. In fact, the comparison is more than an analogy, for space and time become inextricably interwoven in the theory of relativity, united into what physicists call *spacetime*.

Our psychological perception of time differs so radically from the physicist's model that even many physicists have come to doubt whether some vital ingredient has been omitted. Eddington once remarked that there is a sort of 'back door' into our minds through which time enters in addition to its usual route through our laboratory instruments and senses. Our sensation of time is somehow more elementary than our sensation of say, spatial orientation or matter. It is an internal, rather than a bodily experience. Specifically, we feel the *passage* of time — a sensation which is so pronounced that it constitutes the most elementary aspect of our experience. It is a kinetic backdrop against which all our thoughts and activity are perceived.

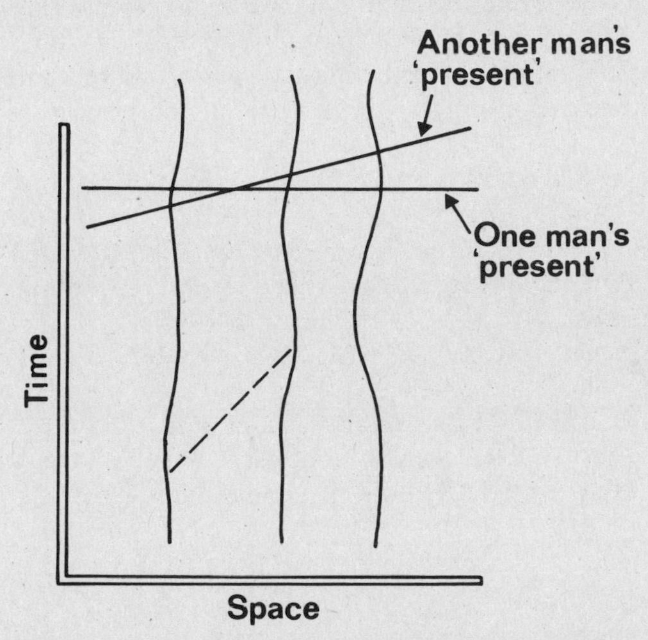

21 Physicists do not regard time as passing but laid out as part of 'space-time', a four-dimensional structure depicted here as a two-dimensional sheet by suppressing two space dimensions. A point on the sheet is an 'event'. The wiggly lines are the paths of bodies that move about; the broken line is the path of a light signal sent between two bodies. The horizontal line through the figure represents a slice through all of space at one instant from the point of view of one observer. Another observer, moving differently, would require the oblique slice. Thus, there must be a temporal (vertical) extension to make sense of the world. There is no universal 'slice' representing a single, common, 'present'. For that reason, division into a universal past, present and future is impossible.

In their search for this mysterious time-flux many scientists have become deeply confused. All physicists recognize that there is a past–future asymmetry in the universe, produced by the operation of the second law of thermodynamics. But when the basis of that law is carefully examined, the asymmetry seems to evaporate.

This paradox can easily be illustrated. Suppose, in a sealed room, the top is removed from a bottle of scent. After a while the scent will have evaporated and dispersed throughout the room, its perfume apparent to anyone. The transition from liquid scent to perfumy air — from order to disorder — is irreversible. We should not expect, however long we wait, for the disseminated scent molecules to find their way back spontaneously into the scent bottle and there to reconstitute the liquid. The evaporation and diffusion of the scent provides a classic example of asymmetry between past and future. If we witnessed a film showing the scent returning to the bottle we should spot immediately that the film was being run backwards in the projector. It is not reversible.

Yet there is a paradox here. The scent evaporates and disperses under the impact of billions of molecular bombardments. The molecules of air in their ceaseless thermal agitation serve to knock the scent molecules about at random, shuffling and reshuffling, until the scent is inextricably mixed with the air. However, any given individual molecular collision is perfectly reversible. Two molecules approach, bounce and retreat. Nothing time asymmetric in that. The reverse process would also be approach, bounce and retreat.

The mystery of time's arrow — how can a past-future asymmetry come from symmetrically colliding molecules — has exercised the

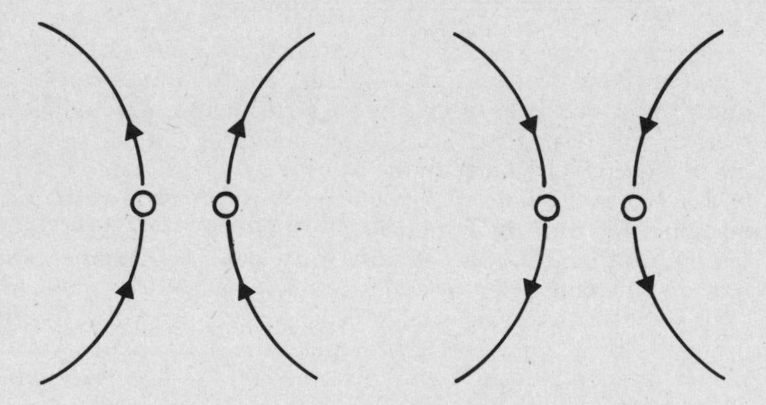

22 The origin of time asymmetry in the world is a mystery when we examine matter at the atomic level. The collision between any two molecules is completely reversible, and displays no preferred past-future orientation.

imaginations of many eminent physicists. The problem was first stated clearly by Ludwig Boltzmann in the late nineteenth century, but the controversy continues today. Some scientists have asserted that there exists a peculiar non-material quality, a time-flux, that is responsible for the arrow of time. They assert that ordinary molecular motions are incapable of imprinting a past–future asymmetry on time, so that this extra ingredient, the time-flux, is essential. Efforts have even been made to trace the origin of the flux to quantum processes or the expansion of the universe. In many ways the belief in a time-flux is closely analogous to — and equally as dubious as — the belief in a life-force.

The mistake is to overlook the fact that time asymmetry, like life, is a holistic concept, and cannot be reduced to the properties of individual molecules. There is no inconsistency between symmetry at the molecular level, and asymmetry on a macroscopic scale. They are simply two different levels of description. One suspects then, that time doesn't really 'flow' at all; it's all in the mind.

When we try to pin down the origin of the time-flux in our perceptions we encounter the same tangle of paradox and confusion that greets attempts to understand the self, and it is hard to resist the impression that the two problems are really closely related. It is only in the flowing river of time that we can perceive ourselves. Hofstadter has written of the 'whirling vortex of self-reference' that produces what we call consciousness and self-awareness, and I strongly believe that it is this very vortex that drives the psychological time-flux. It is for this reason I maintain that the secret of mind will only be solved when we understand the secret of time.

Naïve images of time are to be found everywhere in art and literature: time's arrow, the river of time, time's chariot, time marching on. It is often said that the 'now' or present moment of our consciousness is steadily moving forward through time from past to future, so that, eventually, the year 2000 will become 'now' and by the same token the instant in which you read this will by now have been passed over and consigned to history. Sometimes the now is considered anchored, and time itself is thought to flow, as a river flows past a bankside observer. These images are inseparable from our feelings of free will. The future seems not yet formed, and thus capable of being shaped by our actions before it arrives. Yet surely all this is rubbish?

Problems instantly crop up when one tries to defend the above imagery. A conversation in 1983 between a physicist and a sceptic might go something like this:

Sceptic: I just came across this quote from Einstein: 'You have to accept the idea that subjective time with its emphasis on the now has no objective meaning . . . the distinction between past, present and future is only an illusion, however persistent.' Surely Einstein must have been off his rocker?

Physicist: Not at all. In the external world there is no past, present and future. How could the present ever be determined with instruments? It's a purely psychological concept.

Sceptic: Oh come now, you can't be serious. Everybody knows the future hasn't happened yet, whereas the past is gone — we remember it happening. How can you confuse yesterday with tomorrow, or today for that matter?

Physicist: Of course you must make a distinction between various days in sequence, but it's the labels you use that I object to. Even you would agree that tomorrow never comes.

Sceptic: That's just a play on words. Tomorrow does come, only when it does we call it today.

Physicist: Precisely. Every day is called today on that day. Every moment is called 'now' when it is experienced. Division into *the* past and *the* future is the result of a linguistic muddle. Let me help you sort it out. Each instant of time can be ascribed a definite date. For example, 2 p.m. on 3 October 1997. The dating system is arbitrary, but once we have decided on a convention the date of any particular event or moment is fixed once and for all. By giving date labels to all events, we can describe everything in the world without recourse to dubious constructions like past, present and future.

Sceptic: But 1997 *is* in the future. It hasn't happened yet. Your date system ignores a crucial aspect of time: namely, its flow.

Physicist: What do you mean '1997 *is* in the future'? It is in the past of 1998.

Sceptic: But it's not 1998 *now*.

Physicist: Now?

Sceptic: Yes, *now*.

Physicist: When is now? Every moment is 'now' when we experience it.

Sceptic: *This* now. I mean *this* now.

Physicist: You mean the 1983 now?

Sceptic: If you like.

Physicist: Not the 1998 now?

Sceptic: No.

Physicist: Then all you are saying is that 1997 is in the future of 1983, but in the past of 1998. I don't deny that. It is precisely what my dating system describes. Nothing more. So you see, your talk of the past and future is unnecessary, after all.

Sceptic: But that's absurd! 1997 hasn't happened yet. That is a fact you will surely agree with?

Physicist: Naturally. All you are saying is that our conversation occurs before 1997. Let me repeat. I don't deny there is an ordered sequence of events, with a definite before–after or past–future relation between them. I am simply denying the existence of *the* past, *the* present and *the* future. There clearly is not *a* present, for you and I have both experienced many 'presents' in our life. Some events lie in the past or future of other events, but the events themselves are simply *there*, they don't *happen* one by one.

Sceptic: Is that what some physicists mean when they say that past and future events exist alongside the present — that they are somehow *there*, but we only come across these events one after the other?

Physicist: We don't really 'come across' them at all. Every event of which we are conscious, we experience. They don't lie in wait for us to creep up on them, temporally speaking. There are simply events, and mental states associated with them. You talk as though today's mind is somehow transported forward in time to stumble on tomorrow's events. Your mind is extended in time. Tomorrow's mental states reflect tomorrow's events, today's reflect today's.

Sceptic: Surely my consciousness moves forward from today to tomorrow?

Physicist: No! Your mind *is* conscious both today and tomorrow. Nothing *moves* forward, backward or sideways.

Sceptic: But I *feel* time passing.

Physicist: Hold on a minute, if you will forgive the expression. First you say your mind is moving forward in time, then you say that time itself is moving forward. Which is it to be?

Sceptic: I see time as like a flowing river, bringing future events towards me. Either I can see my consciousness as fixed, and time flowing through it, from future to past, or time is fixed and my consciousness moves from past towards the future. I think the two descriptions are equivalent. The motion is relative.

Physicist: The motion is illusory! How can time move? If it moves it must have a speed. What speed? One day per day? It's nonsense. A day is a day is a day.

Sceptic: But if time doesn't pass, how do things change?

Physicist: Change occurs because objects move about through space *in* time. Time doesn't move. When I was a child I used to wonder 'Why is it *now*, rather than some other time?' When I grew up I learned that the question was meaningless. It can be asked at every moment of time.

Sceptic: I think it's a perfectly legitimate question. After all, why is it 1983?

Physicist: Why is *what* 1983?

Sceptic: Well, why is it 1983 *now*?

Physicist: Your question is a bit like asking 'Why am I *me* and not somebody else?' I am myself by definition — whichever person asks the question. Obviously in 1983 we regard 1983 as 'now'. The same would apply for any year. A legitimate question could be 'Why am I living in 1983 and not, say, 5,000 B.C.?' or 'Why are we having this conversation in 1983 and not 1998?' but there is no need to appeal to notions of past, present and future at all in such discussions.

Sceptic: I'm still not convinced. Almost all our daily thoughts and activities, the tense structure of our language, our hopes, fears and beliefs, are rooted in the fundamental distinction of past, present and future. I am afraid of death, because I have yet to face it, and I am uncertain what lies beyond. But I am not afraid because I don't know of my existence before birth. We can't be afraid of the past. Again, the past is unalterable. We know what happened because of our memories. But we don't know the future, and we believe that it is undetermined, that our actions can change it. As for the present, well, that is our instant of contact with the external world, when our minds can order our bodies to act. Byron wrote 'Act, act in the living present'. That sums it up admirably for me.

Physicist: Most of what you say is true, but still does not require a moving present. Of course there is an asymmetry between past and future, not just in our experiences such as memory, but in the external world. The second law of thermodynamics, for example, ensures that systems tend to become more and more disordered. Other systems possess accumulating records and 'memory'. Think of the craters on the moon: that is a record of past, not future events. All you are saying is that later brain states have more stored information than earlier brain states. We then make the mistake of translating that simple fact into the muddled and ambiguous words 'We remember the past, not the future' in spite of the fact that *the* past is a meaningless phrase. Indeed, in 1998 we shall remember 1997, which *is* in the future of 1983. Stick to dates and you don't need tenses, or the flow of time, or the now.

Sceptic: But you just said 'shall remember' yourself.

Physicist: I could have said: 'My brain state in 1998 records information about events in 1997. But 1997 is in the future of 1983, so is not recorded in my brain states of 1983'. See, no need for past and future after all.

Sceptic: What about the fear of the future, freewill and unpredictability? If the future already exists that must mean complete determinism. Nothing can be changed. Freewill is a sham.

Physicist: The future doesn't 'already' exist. That statement is a contradiction in terms, for it says 'events exist simultaneously with events prior to them' which is obviously nonsense by definition of the word 'prior'. As for unpredictability, that is a practical limitation. It's true we can only predict certain simple events, such as an eclipse of the sun, because of the world's complexity. But predictability is not the same as determinism. You are mixing your epistemology with your metaphysics. Future states of the world could all be determined by prior events, but still be unpredictable in practical terms.

Sceptic: But is the future determined? Sorry. Are all events completely determined by prior events?

Physicist: Actually no. For example, the quantum theory reveals that, at the atomic level, events occur spontaneously, without complete prior causation.

Sceptic: So the future doesn't exist! We *can* change it!

Physicist: The future will be what it will be, whether our actions beforehand are involved or not. The physicist views spacetime as laid out like a map, with time extending along one side. Events are marked as points on the map — some events are linked by causal relations to prior events, others, like the decay of a radioactive nucleus, are labelled 'spontaneous'. It's all *there*, whether the causal links are incorporated or not. So my contention that there is no past, present and future says nothing about freewill or determinism at all. That's quite a separate subject — and a minefield of confusions.

Sceptic: You still haven't explained to me why I *feel* the flow of time.

Physicist: I'm not a neurologist. It has probably got something to do with short-term memory processes.

Sceptic: You're claiming it's all in the mind — an illusion?

Physicist: You would be unwise to appeal to your feelings to attribute physical qualities to the external world. Haven't you ever felt dizzy?

Sceptic: Of course.

Physicist: But you do not attempt to attribute your dizziness to a rotation of the universe, in spite of the fact that you *feel* the world spinning round?

Sceptic: No. It's clearly an illusion.

Physicist: So, I maintain that the whirling of time is like the whirling of space — a sort of temporal dizziness — which is given a false impression of reality by our confused language, with its tense structure and meaningless phrases about the past, present and future.

Sceptic: Tell me more.

Physicist: Not now. I've run out of time . . .

What can one conclude from this sort of exchange? There is no doubt that in organizing our daily affairs we depend heavily upon the concepts of past, present and future, and never question that time really does pass. Even physicists soon lapse back to that way of talking and thinking (as we saw above) once their analytical faculties have been withdrawn. Yet it must be conceded that the closer we scrutinize these concepts the more slippery and ambiguous they seem to become, and all our statements end up as either tautologous or meaningless. The physicist has no need of the flow of time or the now in the world of physics. Indeed, the theory of relativity rules out a universal present for all observers. If there is any meaning at all to these concepts (and many philosophers, such as McTaggart, deny that there is[1]) then it would seem they belong to psychology rather than physics.

This raises an intriguing theological question. Does God experience the passage of time?

Christians believe that God is eternal. The word 'eternal' has, however, been used to mean two rather different things. In the simpler version, eternal means everlasting, or existing without beginning or end for an infinite duration. There are grave objections to such an idea of God, however. A God who is in time is subject to change. But what causes that change? If God is the cause of all existing things (as the cosmological argument of Chapter 3 suggests), then does it make sense to talk about that ultimate cause itself changing?

In the earlier chapters we have seen how time is not simply there, but is itself part of the physical universe. It is 'elastic' and can stretch or shrink according to well-defined mathematical laws which depend on the behaviour of matter. Also, time is closely linked to space, and space and time together express the operation of the gravitational field. In short, time is involved in all the grubby details of physical processes just as much as matter. Time is not a divine quality, but can be altered, physically, even by human manipulation. A God who is in time is, therefore, in some sense caught up in the operation of the physical universe. Indeed, it is quite likely that time will cease to exist at some stage in the future (as we shall see in Chapter 15). In that case God's own position is obviously insecure. Clearly, God cannot be omnipotent if he is subject to the physics of time, nor can he be considered the creator of the universe if he did not create time. In fact, because time and space are inseparable, a God who did not create time, created space neither. But as we have seen, once spacetime existed, the appearance of matter and order in the universe could have occurred automatically as the result of perfectly natural activity. Thus, many would argue that God is not really needed as a creator at all *except* to create time (strictly, spacetime).

So we are led to the other meaning of the word eternal — 'timeless'. The concept of a God beyond time dates at least from Augustine who (as we saw in Chapter 3) suggested that God created time. It has received support from many of the Christian theologians. St. Anselm expresses the idea as follows: 'You [God] exist neither yesterday, today, nor tomorrow, but you exist directly right outside time.'[2]

A timeless God is free of the problems mentioned above, but suffers from the shortcomings already discussed on page 38. He cannot be a personal God who thinks, converses, feels, plans, and so on for these are all temporal activities. It is hard to see how a timeless God can act at all in time (although it has been claimed that this is not impossible). We have also seen how the sense of the existence of the self is intimately associated with the experience of a time-flow. A timeless God could not be considered a 'person' or individual in any sense that we know. Misgivings of this score have led a number of modern theologians to reject this view of an eternal God. Paul Tillich writes: 'If we call God a living God, we affirm that he includes temporality and with this a relation to the modes of time.'[3] The same sentiment is echoed by Karl Barth: 'Without God's complete temporality the content of the Christian message has no shape.'[4]

The physics of time also has interesting implications for the belief that God is omniscient. If God is timeless, he cannot be said to think, for thinking is a temporal activity. But can a timeless being have knowledge? Acquiring knowledge clearly involves time, but knowing as such does not — provided that what is known does not itself change with time. If God knows, for example, the position of every atom today, then that knowledge will change by tomorrow. To know timelessly must therefore involve his knowing all events throughout time.

There is thus a grave and fundamental difficulty in reconciling all the traditional attributes of God. Modern physics, with its discovery of the mutability of time, drives a wedge between God's omnipotence and the existence of his personality. It is difficult to argue that God can have both these qualities.

10. Free will and determinism

> 'Nothing would be uncertain and the future, as the past, would be present to [our] eyes.'
>
> Pierre de Laplace

When Newton invented his laws of mechanics, many people took this to be the death of the freewill concept. According to Newton's theory, the universe is like a giant clockwork, unwinding along a rigid, predetermined pathway towards an unalterable final state. The course of every atom is presumed to be legislated and decided in advance, laid down since the beginning of time. Human beings were seen as nothing but component machines caught up irresistably in this colossal cosmic mechanism. Then along came the new physics with its relativity of time and space and its quantum uncertainty. The whole issue of freedom of choice and determinism went back into the melting pot.

There seems to be a fundamental antagonism between the two theories that constitute the foundations of the new physics. On the one hand, the quantum theory endows the observer with a vital role in the nature of physical reality; as we have seen, many physicists claim that there is concrete experimental evidence against the notion of 'objective reality'. This appears to offer human beings a unique ability to influence the structure of the physical universe in a way that was undreamt of in Newton's day. On the other hand, the theory of relativity, which demolishes the concept of a universal time, and an absolute past, present and future, conjures up a picture of a future that in some sense already exists, and so cuts from under our feet the victory won with the help of the quantum factor. If the future is *there*, does it not mean that we are powerless to alter it?

In the old Newtonian theory, every atom moves along a trajectory that is uniquely determined by the forces which act on it. The forces in turn are determined by other atoms, and so on. Newtonian mechanics permits, in principle, the accurate prediction of everything that will ever happen on the basis of what can be known at one instant. There is a rigid network of cause and effect, and every phenomenon, from the tiniest jiggle of a molecule to the explosion of a galaxy, is determined in detail long in advance. It was this conception of mechanics that led Pierre de Laplace (1749–1827) to declare that if a being knew at one instant the positions and motions of every particle in the universe he would have at his disposal all the information necessary to compute the entire past and future history of the universe.

The argument of the 'Laplacian calculator' is not as open and shut as it might seem, however. First, there is the problem about whether a brain can, even in principle, compute its own future state. MacKay has argued that, for each individual, complete *self*-predictability is impossible, even in a mechanistic universe of the Newtonian variety.[1] For suppose a super-scientist could peer into your brain and compute precisely what you will do on some future occasion, this does not logically preclude free will in a certain sense. The reason is that, though he may be correct in his prediction, he cannot tell you of that prediction (before the event) without messing up his calculation. When he tells you, for example, 'Yes, you will clap your hands' your brain state is inevitably altered from what it was before he told you the prediction; altered, that is, by this new piece of information. You would then have no reason to believe the prediction, since it was based on a now-altered brain state. Hence, no prediction can be made *that you would be correct in believing* about your future behaviour. MacKay thus argues that, however predictable and inevitable your behaviour may be to a hypothetical super-scientist who withholds that prediction, it remains logically unpredictable for you and so preserves at least an element of what is normally understood as free will.

Then there is the question of whether the universe is predictable after all in Newtonian mechanics. Recent advances in the mathematical description of mechanical systems have revealed that some types of forces are responsible for such acute instability in the evolution of certain systems that predictability is a meaningless concept. Whereas in a 'normal' mechanical system, slight variations of the initial conditions produce only slightly altered behaviour, these ultra-sensitive systems will evolve in totally different ways from two initial states that differ from each other by only an infinitesimal amount.

Furthermore, the discoveries of modern cosmology reveal that our universe should have an expanding horizon in space, and that every day new disturbances and influences cross into the universe from the regions beyond the horizon. Because these regions have never been in causal communication with our part of the universe since the beginning of time, it is not possible, even in principle, for us to know what these incoming influences might be.

The most important argument, however, against complete predictability is the quantum factor. According to the basic principles of the quantum theory, nature is inherently unpredictable. Heisenberg's famous uncertainty principle assures us that there is always an irreducible indeterminism in the operation of subatomic systems. In the microworld, events occur that have no well-defined cause.

Does not the collapse of determinism conflict with the theory of relativity? In this theory there is no universal present, and the entire past and future of the universe are regarded as existing as an indivisible whole. The world is four-dimensional (three of space, one of time), and all events are simply *there*: the future does not 'happen' or 'unfold'.

Any conflict is, in fact, illusory. Determinism concerns the question of whether every event is completely determined by a prior cause. It says nothing about whether that event is *there*. After all, the future will be what it will be regardless of whether it is determined by prior events or not. The four-dimensional perspective of relativity simply forbids us to slice up spacetime, in any absolute way, into universal instants of time. The notion of two events in different places being 'simultaneous' is relative to one's state of motion. They may be judged to occur at the same moment by one observer, but one after the other by another observer. We must therefore regard the universe as extended in time as well as space. But the theory tells us nothing about whether the temporal extension includes rigid links of cause and effect between the events there displayed. So in spite of the fact that past, present and future seem to have no objective meaning, the theory of relativity does not forbid a human being from deciding later events by his earlier actions. (Recall that the earlier–later ordering relation *is* an objective property of time, even though *the* past and *the* future are not.)

However, it is not at all clear that an indeterministic universe is, in fact, what is wanted to establish free will. Indeed, the determinist would argue that freewill is only possible in a *deterministic* universe. A free agent is, after all, one who is able to cause certain acts in the physical world. In an indeterministic universe, events occur that are uncaused. But can you be responsible for your acts unless they are caused — caused by *you*? Proponents of free will assert that the activities of a person are *determined*, for instance, by his character, inclinations, personality.

Suppose a docile and peaceful man were suddenly to commit an act of violence. The indeterminist could say, 'It was a spontaneous event, with no prior cause. You cannot blame the man.' The determinist, on the other hand, would declare the man to be responsible, but take comfort in the fact that he could be rehabilitated by education, persuasion, psychotherapy, drugs, and so on which would *cause* him to act differently in the future. Indeed, a central message of most religious thinking is that we are able to improve our characters. But that is only possible to the extent to which our future characters are determined by our earlier decisions and actions. It is important to realize that determinism does not imply events occur *in spite of* our actions. Some events occur because *we* determine them.

Determinism must not be confused with the doctrine of fatalism, which asserts that future events are entirely beyond our control. 'It is all written in the stars', declares the fatalist. 'What will be will be.' The soldier who behaves recklessly on the battlefield in the face of a hail of bullets while thinking 'if my number is on it, no precaution will avert death' is a fatalist. Some Oriental religions contain fatalist overtones, and many people are inclined to lapse into fatalism from time to time, especially as far as major world affairs are concerned. 'It is beyond my power to influence events, one way or the other.' That is doubtlessly true. Ordinary people cannot avert world war or prevent the devastation of a city by the impact of a huge meteor. Yet in daily life we continually influence the outcome of events in countless small ways. Nobody would seriously say, 'Why bother to look when I cross the road, for my fate is already decided.'

Still, we have strong misgivings about determinism, which is why so many people are relieved that the quantum factor apparently demolishes the idea. Certainly our desire for freedom includes the requirement that what we decide may actually be caused by us to happen. But in a completely deterministic universe the decision is *itself* predetermined. In such a universe, though we may perhaps do as we please, *what* we please is beyond our control. The argument goes like this. When you choose to drink tea rather than coffee, the decision is due to environmental influences (such as, tea is cheaper), physiological factors (coffee is a stronger stimulant), cultural dispositions (tea is a traditional drink), and so on. Determinism asserts that every decision — every whim — is determined in advance. If that is so, however free you may feel to choose tea or coffee, in reality your choice was destined from the moment you were born — even before. In a fully deterministic universe *everything* is determined from the instant of the creation. Does this make us less than free?

The problem is that it is very hard to decide exactly what sort of freedom we want. One suggestion is that *real* freedom to choose tea or coffee means that if the circumstances leading up to the choice were repeated, with everything in the universe exactly the same (including your brain state, because your brain is part of the universe) then there is a probability that you would choose differently on the repeat performance. Such an outcome is clearly incompatible with determinism. But how could this ultimate version of freedom ever be tested? How could the universe ever be reconstituted in identical form? If that is what is meant by freedom its existence must be a matter of pure faith.

Perhaps freedom means something else: unpredictability in the MacKay sense? What you will do *is* determined by elements beyond your control, but you can never know, even in principle, *what* it is that you will do. Is this enough to satisfy the desire for free will?

Another view of freedom is that some (or all) events are caused, but that the events caused by us have no cause from within the natural universe. Specifically, this idea asserts that our *minds* are external to the physical world (the dualist philosophy), but they can somehow reach into it and influence what happens. Thus, as far as the physical world alone is concerned not all events can be determined, because mind is not part of the physical world. One can still ask, what causes the mind to decide in the way it does? If those causes originate in the physical world (and clearly some do) then we are back with determinism, and the introduction of a non-physical mind is an empty embellishment. But if some of those causes are non-physical, does that make us more free? If we have no control over the non-physical causes, then we are no better off than we are with uncontrollable physical causes. But if we can control the causes of our own decisions, what determines how we choose to exercise that control: more external influences (physical or non-physical), or *us*? 'I do it because I make myself make myself make myself . . .' Where does the chain end? Must we fall into an infininte regress? Can we say that the first link in the chain is *self*-caused: it requires no cause from outside itself? Does this concept of self-causation — causeless causes — have any meaning?

So far we have been ignoring indeterminism. Most physicists would claim that the conflict between determinism and free will is irrelevant because we know that the quantum factor disproves determinism anyway. But we must be careful here. Quantum effects are probably too small to have much influence on the operation of the brain at the neuron level, but if they did we would surely have not free will, but breakdown. A quantum fluctuation that forced a neuron to fire when it would normally not (or vice versa) is surely to be regarded as an interference to the otherwise normal operation of the brain. If electrodes were planted in your brain and triggered at random by an external source, you would regard that form of interference as a *reduction* of your freedom: someone 'taking over', or at least impeding, the operation of your brain. How can random quantum quirks inside your head represent anything other than 'noise'? You decide to raise your arm, the neurons fire in the correct sequence, but a quantum fluctuation disturbs the signal. Your leg moves instead. Is that freedom? That is the fundamental problem of indeterminism: your actions may not be under your control because they are not *determined*, by you or anything else.

Still, it is hard to resist the impression that the quantum factor does hold out some hope for freedom. Certainly we do not wish the sequence of neuron firings to be interrupted once it has been initiated, but it might be argued that quantum effects are only important at the first stage — the initiation. Imagine a neuron that is primed to fire, and needs only the slightest disturbance at the atomic level to trigger it. The quantum theory says there is a definite probability that the neuron will or will not fire. The actual outcome is undetermined. This is where the mind (or soul) comes in. It says (subconsciously) 'Electron move to the right!', or some such command, and the neuron fires. No violation of a physical law is involved in this version of mind-over-matter, because there was a distinct chance that the neuron would have fired anyway. The mind simply tipped the balance of odds to make sure it did.

Unfortunately, however, quite apart from the lack of any evidence that the brain really is so delicately balanced (and if it were, extraneous electric and magnetic disturbances might then swamp the effect of the mind) this scenario runs into the problem already discussed above — the question of what causes the mind to command the electron to move to the right in the first place. It also runs into strong objection from those who reject the dualist solution of the mind–body problem, for they would maintain that the mind is not a substance capable of *acting on* the brain anyway. If the mind is regarded as the software representative of the brain's electrochemical structure, to talk of the mind acting *on* the brain is to fall for a confusion of levels once again. It is as meaningless as attributing the publication of a novel to one of its characters, or saying that a switching circuit in a computer fires because the program forces it to.

None of the foregoing really gets to grips with the central paradox of the quantum theory, which is the unique role played by the mind in determining reality. As we have seen, the act of observation causes the ghostlike superposition of potential realities to cohere into a single, concrete reality. Left to its own devices, an atom cannot make a choice. We have to observe it before a particular outcome is realized. The fact that you can decide to create either an atom-at-a-place or an atom-with-a-speed confirms that, whatever its nature, your mind does, in a sense, reach into the physical world. But now we can once again ask *why* you decided to measure, say, the position rather than the motion of the atom. Is this freedom to construct reality any more powerful than the already existing freedom to influence the external world by moving objects around, say, by touch?

Many physicists these days are inclining towards the so-called Everett many-universes interpretation of the quantum theory. This view (briefly discussed in Chapter 8) has bizarre implications for the subject of free will. According to Everett, every possible world is actually realized, with all the alternative worlds coexisting in parallel. This duplication of worlds extends to human choices. Suppose you are faced with a choice — tea or coffee? The Everett interpretation says that the universe immediately divides into two branches. In one of the branches you have tea, in the other coffee. This way you have everything!

The many-universes theory would seem to satisfy the ultimate criterion for freedom of choice discussed above. When the split occurs, the circumstances leading to each outcome are truly identical in all respects — they are in fact the *same* universe — yet two different choices are made. (As noted before, no one can directly verify this theory, for any*one* must be restricted to one branch of the dividing universe.) Yet the victory seems a pyrrhic one. If you can't avoid making *all* possible choices, are you really free? The freedom seems overdone, destroyed by its own success. You want to choose tea *or* coffee, not tea *and* coffee.

But now the many-universes proponent says: 'Ah! But what do you mean by *you* here?' The 'you' that actually has the tea is not the same as the 'you' that has the coffee. They inhabit different universes. If nothing else, these two individuals we lightly referred to as 'you' will differ in their perceptual experience (for instance, in the taste of the drink). They cannot be the *same* person. So, offered the choice, you don't actually have tea and coffee after all. Whichever of the two product 'yous' one is discussing, *that* you has made its choice. According to this view, then, saying that you have chosen tea in preference to coffee amounts to no more than a definition of 'you'. To say 'I chose tea' means, simply, 'I am the tea-drinker'. Thus, although a single 'you' was faced with the choice, the outcome involved two individuals, not one. In the Everett theory the self is continually multiplied into countless near copies. (The implications of this for the traditional concept of a distinct soul would be interesting to explore.)

Much has been written about the relationships between free will and the question of blame and responsibility for crime. If free will is illusory, why should anyone be blamed for their acts? And if all is predetermined, every one of us is locked into a course of action that is decided in advance of our existence. In an Everett multi-universe, could not the felon plea that at least one component of his multi-self is obliged by the laws of the quantum theory to commit the crime? We must, however, turn aside from this minefield and ask about the position of God in a deterministic universe. As soon as God is injected into the picture we bring down upon ourselves a deluge of puzzles.

Can God exercise free will and make decisions?

If man possesses free will, surely God does too? In which case many of the foregoing problems concerning the freedom concept extend to God. In addition we have all the usual perplexities associated with an infinite and omnipotent Deity. If God has a *plan* for the universe, which is implemented as part of his will, why does he not simply create a deterministic universe in which the goal of the plan is inevitable? Or better still create it with the plan achieved? If the universe is indeterministic, however, does that not mean that God's power is limited because of his inability to predict or decide what the outcome will be?

It could perhaps be argued that God is free to relinquish some of his power if he wishes. He can give *us* free will to act against his plan if we so desire, and he can give atoms the quantum factor to turn his creation into a cosmic game of chance. But there is a logical problem of whether a truly omnipotent agent can relinquish some power.

The notion of freedom implied by omnipotence is quite different from the sort of freedom that humans enjoy. You may be free to choose tea or coffee, but only so long as supplies exist. You are not free to do *anything* you please — to swim the Atlantic or turn the moon to blood, for example. Human power is limited, and only a small range of desires are capable of being fulfilled. By contrast, the power of an omnipotent God is without limit, and such a being is free to have whatever he chooses.

Omnipotence raises some awkward theological questions. Is God free to prevent evil? If he is omnipotent, yes. Why then does he fail to do so? This devastating argument was deployed by David Hume: if the evil in the world is from the intention of the Deity, then he is not benevolent. If the evil is contrary to his intention, he is not omnipotent. He cannot be both omnipotent and benevolent (as most religions claim).

One response to this argument is that evil is due entirely to human activities; because God has given us freedom, we are free to do evil and thus frustrate God's plan. Still, if God is also free to prevent us from doing evil must he not share some of the responsibility if he fails to do so? When a parent allows an unruly child to run amok, attacking neighbours and causing damage, we would normally lay a portion of the blame at the parent's feet. Must we therefore conclude that evil (in perhaps a limited amount) is all part of God's plan? Or is God not free after all to prevent us from acting against him?

Fresh puzzles crop up if the Christian doctrine is followed in which God is believed to transcend time, for the concept of freedom to choose is intrinsically a temporal one. What meaning would it have to make a choice, not at a particular moment, but timelessly? And if God already knows the future, what meaning can we attach to a cosmic plan and our own participation in it? An infinite God will know what is happening everywhere. But as we have seen, there is no universal present moment, so God's knowledge *must* extend in time if it extends in space. So we conclude that it is meaningless for a Christian eternal God to have freedom of choice. But can we believe that man possesses a faculty not available to his creator? We seem forced to the paradoxical conclusion that freedom of choice is actually a *restriction* that we suffer — namely, our inability to know the future. God, released from the prison of the present, has no need of free will.

The problems seem insurmountable. The new physics undoubtedly gives a new slant to the longstanding enigma of free will and determinism, but it does not solve it. The quantum theory undermines determinism, but brings its own crop of difficulties concerning freedom, not least of which is the possibility of multiple realities. The theory of relativity offers us a universe extended in time as well as space, but still leaves the door open for some sort of freedom of action. No doubt future developments in our understanding of time will cast new light on these fundamental problems of our existence.

11. The fundamental structure of matter

> 'By getting to smaller and smaller units, we do not come to fundamental units, or indivisible units, but we *do* come to a point where division has no meaning.'
>
> Werner Heisenberg

> 'The present day attempts at unified field theory are really enormously simple.'
>
> I.M. Singer

Science is possible only because we live in an ordered universe which complies with simple mathematical laws. The job of the scientist is to study, catalogue and relate the orderliness in nature, not to question its origin. But theologians have long argued that the order in the physical world is evidence for God. If this is true, then science and religion acquire a common purpose in revealing God's work. Indeed, it has been argued that the emergence of Western scientific culture was actually stimulated by the Christian–Judaic tradition, with its emphasis on God's intentional organization of the cosmos — an organization which could be discerned by the use of rational scientific enquiry. It is a philosophy that seems to be captured in the following lines by Stephen Hales (1677–1761):

> Since we are assured that the all-wise Creator has observed the most exact proportions, of number, weight and measure, in the make of all things, the most likely way therefore, to get any insight into the nature of those parts of the creation, which come within our observation, must in all reason be to number, weigh and measure.

That the universe is ordered seems self-evident. Everywhere we look, from the far-flung galaxies to the deepest recesses of the atom, we encounter regularity and intricate organization. We do not observe matter or energy to be distributed chaotically. They are arranged instead in a hierarchy of structure: atoms and molecules, crystals, living things, planetary systems, star clusters, and so on. Moreover, the behaviour of physical systems is not haphazard, but lawful and systematic. Scientists frequently experience a sense of awe and wonder at the subtle beauty and elegance of nature.

It is helpful to distinguish between different sorts of order. First there is the order of simplicity, seen for example in the regularities of the solar system or the periodic oscillations of a pendulum. Then there is the order of complexity, such as the arrangement of gases in the swirling atmosphere of Jupiter, or the complex organization of a living creature. The distinction is another example of reductionism versus holism. Reductionism seeks to uncover simple elements within complex structures, while holism directs attention to the complexity as a whole. The order of complexity suggests to many an element of purpose, in which all the component parts of a system fit together harmoniously in a cooperative way to achieve some particular end. In this chapter we shall look at the order of simplicity and see how very recent discoveries in fundamental physics confirm that mathematical regularity controls the vital processes of nature. We shall return to explore the order of complexity in the following chapter.

It was proposed by Kant that the human mind inevitably imposes order on the world so as to make sense of it, but I don't believe many scientists are impressed by this argument. Kant knew nothing of atomic or nuclear structure, for example, yet the study of the atom revealed the same sort of mathematical regularities that occur in the organization of the solar system. This is surely a surprising fact and has nothing to do with the way we choose to perceive the world. Moreover, we shall see that subnuclear matter is subject to some simple and powerful symmetry principles. It is difficult to be convinced that, for example, the left and right-hand symmetry in the operation of some of the fundamental forces is of no significance except as a tribute to the tidy nature of the human mind.

The order of simplicity in nature has traditionally been exposed by following scientific reductionism: breaking complex systems into their simpler component parts and studying the components in isolation. The idea that all matter is built out of a small number of basic units — the original 'atoms' — dates from classical Greece, but it is only in this century that technology has advanced to the point where atomic processes may be studied and understood in detail. One of the earliest discoveries, principally due to the work of Lord Rutherford just after the turn of the century, was that atoms are not elementary particles at all, but composite structures with internal parts. Most of the atomic mass is concentrated in a tiny nucleus, only a thousand-billionth of a centimetre in size. The nucleus is surrounded by a cloud of lighter particles — the electrons — extending out to a distance of perhaps a hundred-millionth of a centimetre. Thus, by far the greater part of the atom is empty space. Add to this the fact that the quantum factor precludes precise orbital paths for the electrons, and the atom begins to seem a rather insubstantial and nebulous sort of entity.

The electrons are bound to the nucleus by electric forces. The nucleus is positively charged, and surrounded by an electric field which traps the negatively charged electrons. Long ago it was found that the nucleus is itself a composite body, consisting of two types of particles: protons, which carry the positive charge, and electrically neutral particles called neutrons. Protons and neutrons are both about 1800 times heavier than electrons.

Once this basic architecture was discerned, physicists were able to apply the quantum theory to the atom and thereby reveal a remarkable form of harmony. The essentially wavelike nature of electrons manifests itself by the existence of certain definite 'stationary states' or 'energy levels' in which the electrons reside. Transitions may occur between the levels if energy is absorbed or emitted in the form of photons (packets of light energy). The existence of the levels therefore shows up in the energy of the light, which is deduced from its frequency (colour). An analysis of the light emitted or absorbed by atoms therefore reveals a spectrum of colours in the form of a series of discrete frequencies or spectral lines. The simplest atom is hydrogen, consisting of one proton (the nucleus) and one electron. Its energy levels are given by a simple formula:

$$\frac{1}{n^2} - \frac{1}{m^2}$$

multiplied by a fixed unit of energy, where n and m are whole numbers: 1, 2, 3 . . . Such compact arithmetical expressions are reminiscent of musical tones — the harmonic notes on a guitar string or organ pipe for example — which are also described by simple numerical relations. This is no coincidence. The arrangement of energy levels in an atom is a response to the quantum wave vibrations, just as the frequencies of a musical instrument are a response to sound vibrations.

The atomic harmony would not be so elegant, however, if it were not for the fact that the force which binds the electron to the proton in the hydrogen atom is also mathematically simple. Indeed, the very existence of atoms depends on it. This electric attraction satisfies a famous law of physics known as the inverse square law. It means that if the separation between the proton and electron is doubled the force falls to one quarter of its value; if it is trebled, the force is one-ninth, and so on. It is a tidy mathematical regularity that is also found in the force of gravity: for example, the attraction between the planets and the sun. In that case the inverse square law leads to the celebrated regularities of the solar system, manifested in the computational formulae that predict eclipses and other celestial phenomena. In the atomic case the regularities are of a quantum nature: the arrangement of energy levels and the frequency spectrum of emitted light. But both derive from the simplicity of the inverse square law.

As soon as the structure of the nucleus was apparent, physicists began to wonder about the internal nuclear forces that hold it together. Gravity is too weak, and electric forces are repulsive between like charges, so there is a puzzle about how the protons, each with their positive charge, avoid being thrust apart. Clearly there must exist a strong attractive force to overcome this electric repulsion. Experiments revealed that the nuclear force is much stronger than electricity, and fades away abruptly outside a definite range or distance from the proton. This range is very short — less than the size of the nucleus — so only the nearest neighbour particles are held in its grip. Both neutrons and protons experience the nuclear force. Because the force is so strong, it requires great energy to break most nuclei apart, but this can be done. Heavy nuclei are less stable, and can fission easily, with a resulting energy release.

The nuclear particles are also arranged in discrete quantum energy levels, but here the simplicity of atomic harmony is absent. The nucleus is a complicated structure, not only because of the multiplicity of particles, but also because the nuclear force is not of the simple inverse square form.

As physicists studied the nuclear force in the 1930s in the context of the quantum theory, it became apparent that the nature of force is inseparable from the structure of particles. In daily experience we think of matter and force as quite distinct concepts. Forces can act between material bodies via gravity or electromagnetic effects, or directly through physical contact, but matter is regarded only as the *source* of the force, not as the agency for its transmission. Thus, the sun exerts a gravitational action on the Earth across empty space, and this may be described in the language of fields: the sun's gravitational field, which is otherwise invisible and intangible, interacts with the Earth and exerts a force.

In the subatomic domain, where quantum effects are important, the language and description changes profoundly. It is a central feature of the quantum theory that energy is transmitted in discrete lumps or quanta, which give the theory its name. So, for example, photons are quanta of the electromagnetic field. When two electric particles approach each other, they come under the influence of their mutual electromagnetic fields, and forces operate between them. The forces cause the particles to deviate in their motion. But the disturbance

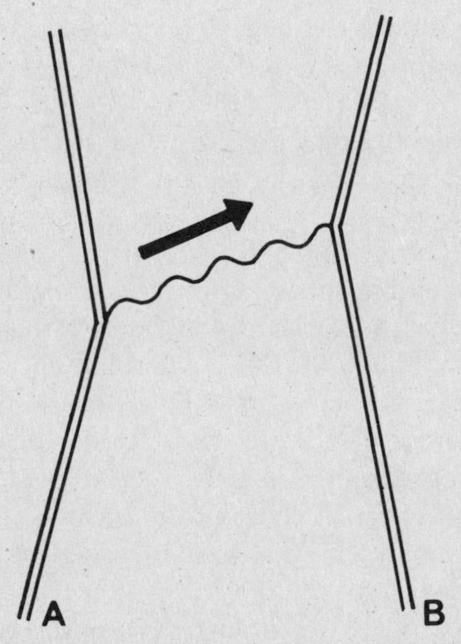

23 At the quantum level, the electromagnetic forces between charged particles A and B are understood in terms of the exchange or transfer of photons. The path of particle A shows a recoil as the photon is emitted. Later, B is deflected by the absorption. In this way, forces between particles are transmitted by other particles (photons in this case). Actually, this description is rather simplified. The transmission involves a complex web of short-lived (or 'virtual') particles travelling in both directions and also buzzing around each of A and B individually. A similar sort of description may be given for nature's other fundamental forces. The diagrams are a symbolic representation of abstract mathematical terms that can be used to compute subatomic processes correctly to very high accuracy.

which one particle inflicts on the other through the field must be transmitted in the form of photons. Therefore, rather than a continuous process, the interaction between charged particles is best envisaged as a sudden impulse due to the transfer of one or more photons.

It is helpful to represent such processes by the use of diagrams invented by Richard Feynman. Figure 23 shows a single photon being transferred between two electrons, which scatter apart as a consequence. This mechanism of interaction has been compared to two tennis players whose behaviour is coupled via the exchange of the ball. Photons, therefore, act rather like messengers, hopping back and forth between charged particles telling them that the other charged particle is there, and inducing a response. Using these ideas, physicists can calculate the effects of many electromagnetic processes at the atomic level. In all cases the experimental results agree with the calculated predictions to stunning accuracy.

The quantum theory of the electromagnetic field was so successful that it was natural for physicists of the 1930s to apply it to the nuclear force field too. This was done by the Japanese physicist Hideki Yukawa, who discovered that the force between protons and neutrons could indeed be modelled by the exchange of messenger quanta, but quanta of quite a different nature from the familiar photons. To reproduce the effects of a very short-ranged force, Yukawa's quanta had to carry mass.

This is a subtle, but important point. The mass of a particle is a measure of its inertia, or resistance to change in motion. A light particle is more easily moved by a given force than a heavy one. If a particle becomes exceedingly light it will be accelerated by any stray forces, and so will tend to travel very fast. In the limiting case that the mass dwindles away to nothing, the particle will always travel at the fastest possible speed, which is the speed of light. This is the case with photons, which can be regarded as massless particles. Yukawa's quanta, on the other hand, have mass, and travel slower than light. Yukawa called them mesons, but they are now known as pions.

Inside the nucleus, pions flit back and forth between the neutrons and protons, gluing them together with nuclear force. Normally they go unseen, because no sooner are they created than they are absorbed again by another nuclear particle. However, if energy is pumped into the system, a pion can fly out to be studied in isolation. This happens when two protons collide at high speed (a process mentioned briefly in Chapter 3). When pions were first discovered this way shortly after the

Second World War, it was a brilliant verification of Yukawa's theory, and the discovery of the pion was hailed as a triumph of theoretical physics in general and the quantum theory of fields in particular. Another distinctive feature of pions is that they are violently unstable and almost instantly decay into lighter particles. One of the decaying products, called the muon, is identical to the electron in all respects except mass. It is considerably heavier than the electron, and also rapidly decays.

Once physicists realized that they could make entirely new splinters of matter by high speed subatomic particle collisions they began to build huge accelerating machines to do the job. These machines are capable of boosting all manner of subatomic fragments to near the speed of light, and the shock of near-luminal impact bursts open a whole new world of subnuclear activity. Once these machines became available, dozens of new and hitherto unsuspected particles appeared. So prolific were the newcomers that physicists rapidly ran out of names for them. For a while the different species of particles resembled a disorganized zoo. Then gradually the physicists' bewilderment subsided as they began to recognize some semblance of order amid the subnuclear debris. Patterns began to emerge.

It has been known since the 1930s that there is not one nuclear force, but two. The strong force glues the nuclear particles together, but there is also a very much weaker force. The weak force is responsible for causing some of the unstable nuclear particles to decay: for example, pions and muons disintegrate under the effect of the weak force. Some particles feel both the strong and weak forces, but others do not feel the strong force. These latter particles tend to be the lighter ones, and include muons, electrons and neutrinos. There are at least two varieties of neutrinos, and they are the most elusive objects known to science. So weakly do they interact with other matter that they could penetrate several light years of solid lead!

The light, weakly interacting particles are given the collective name of leptons. Charged leptons, such as electrons, feel both the weak and electromagnetic forces, but the uncharged neutrinos are blind to electromagnetism. The heavier, strongly interacting particles are called hadrons, and they divide into two families. In one family are the protons and neutrons, together with numerous heavier particles that decay into them. These are known as baryons. The rest are called mesons, and include the pions.

Within these broad family groupings many subgroups can be discerned. The members of any particular subgroup possess properties, such as mass, electric charge and other more technical qualities, that vary systematically from one member to the next. In the 1960s theorists discovered that these step by step systematic properties could be represented in a very elegant way using a branch of mathematics known as the theory of groups. The underlying principle here is the concept of symmetry, and it is probably true to say that once the subatomic symmetry idea finally dawned on the physics community they never looked back.

It has always been appreciated that symmetry plays a vital role in the organization of the natural world. Examples such as the spherical figure of the sun or the regularity of a snowflake or a crystal are familiar to us all. Not all symmetries, however, are geometrical. The symmetry between male and female, or positive and negative electric charge are also useful concepts, but here the symmetries are of an abstract nature. So, too, amid the collections of baryons and mesons, abstract symmetries were discovered which suggested that the members of any particular grouping are closely connected by a simple mathematical scheme. Some flavour of these ideas can be given by analogy with familiar geometrical symmetries. Everyone knows that in a mirror a left hand is reflected into a right hand. Left and right hands form a two-component symmetric system: two reflections in sequence bring you back to the original. There is a sense in which a proton and neutron can be considered similar to a left hand and a right hand. Under 'reflection' a neutron changes into a proton or vice versa. The reflection is not, of course, an ordinary reflection in real space, but is a sort of abstract reflection in an imaginary space (known in the trade as isotopic spin space). Although the symmetry is abstract, its mathematical description is identical to a geometrical symmetry, and the manifestation of that symmetry is real enough. It shows up in the properties of protons and neutrons in scattering experiments, and the way in which they engage the attentions of other particles.

More complex symmetry groups enable a unified description of larger families of particles than just the proton and neutron. Some of these families contain eight, ten or more particles. Occasionally certain symmetries are not apparent at first sight because they are masked by complicating effects, but the combined effort of mathematical analysis and careful experimentation can expose them.

Few physicists fail to be struck by the subtle elegance that these abstract symmetries betray about the inner workings of matter. The entire subnuclear enterprise is founded on the tenacious belief that simplicity lies somewhere at the heart of all natural complexity. Yuval

Ne'eman and Murray Gell-Mann, who were the first to discover hidden symmetry in a collection of eight mesons, called their new principle 'the eightfold way' after a pronouncement by the Buddha: 'This Ariyan Eightfold Path; that is to say: Right view, right aim, right speech, right action, right living, right effort, right mindfulness, right contemplation.'

As more and more symmetries were revealed, particle physicists became deeply impressed by these subtle regularities — regularities that had remained secret, buried in the depths of the atom, since time immemorial. Now, for the first time, they were being witnessed by human beings with the aid of dazzling instruments of advanced technology.

It was not long before physicists began to wonder about the meaning behind the symmetries: 'It seemed as though nature was trying to tell us something,' remarked a leading theorist. At this point the power of mathematical analysis surfaced again. The theory of groups suggested a natural origin for all the family symmetries in terms of a single, underlying, master symmetry. The higher symmetries, it turned out, could all be built out of combinations of a very simple arrangement. Translated into the language of particles, the mathematics suggested that the hadrons were not fundamental at all but, yet again, composites of much smaller particles.

Wheels within wheels! Atoms are made of nuclei and electrons; nuclei are made of protons and neutrons; protons and neutrons are made of. . .? The new building blocks, three levels down from atoms, needed a name. Gell-Mann coined the word quark, and it stuck. Hadrons are built of quarks. The great principle of the ancient Greeks that all matter is constructed from a small number of truly elementary particles (their 'atoms') has proved a tortuous path to follow. Does the buck stop here, or are quarks also composite objects? We shall return to this question shortly.

Quarks stick together in one of two configurations: doublets and triplets. The union of two quarks makes a meson, that of three quarks a baryon. Quarks also reside in quantum energy levels, and can become excited to higher levels by eating energy. Excited hadrons look like other hadrons, and many previously distinct particles are now seen to be just excited states of a single quark combination.

To account for all known hadrons, it is necessary to suppose that there is more than one sort of quark. In the early 1970s the job could be done with three quark 'flavours', as they are whimsically called: 'up', 'down' and 'strange'. Then more hadrons turned up and a fourth quark was added: the 'charmed' quark. Recently still more particles have appeared and two further quarks, 'top' and 'bottom', are deemed necessary. But the success of the quark scheme is astonishing. A huge variety of particle processes can now be understood in a systematic way by appeal to detailed quark calculations.

The underlying assumption of the quark theory is that the quarks themselves are truly structureless, fundamental particles — point-like objects with no internal parts. In this respect they are like the leptons, which are not built of quarks but seem to be fundamental in their own right. In fact there is a natural head-to-head correspondence between quarks and leptons that provides a curious insight into the workings of nature. The link is shown schematically in Table 1. In the right-hand column are the quark flavours, on the left are all the known leptons. Recall that leptons feel the weak force, quarks feel the strong force. Another difference is that leptons have either no electric charge or one unit of charge, whereas quarks have charge of either one-third or two-thirds of a unit.

Table 1

	LEPTONS		QUARKS	
	name	charge	name	charge
I	electron (e)	−1	up (u)	$+\frac{2}{3}$
	electron-neutrino (ν_e)	0	down (d)	$-\frac{1}{3}$
II	muon (μ)	−1	strange (s)	$-\frac{1}{3}$
	muon-neutrino (ν_μ)	0	charmed (c)	$+\frac{2}{3}$
III	tau (τ)	−1	top (t)	$+\frac{2}{3}$
	tau-neutrino (ν_τ)	0	bottom (b)	$-\frac{2}{3}$
	?	?	?	?

Subatomic particles can be divided into two broad classes: leptons and quarks. Quarks are not found individually, but united in groups of two or three; they have fractional electric charge. All ordinary matter is made out of Level I particles. Levels II and III seem to be simple replications of Level I, the particles concerned being highly unstable. There may be further levels as yet undiscovered.

Omitted from this scheme are the messenger particles: the photons, gravitons, gluons, and the mediators of the weak nuclear force, known as W and Z.

In spite of these differences, deeper mathematical symmetries exist that connect quarks and leptons level by level in the table. The first level contains just four particles: the up and down quarks, the electron and its neutrino. Curiously, all ordinary matter is made out of these four particles alone. Protons and neutrons are made from up and down quarks uniting in triplets, while electrons make up the only other subatomic particle needed. Neutrinos just go off into the universe and play no part in the gross structure of matter. As far as we can tell, if all the other particles suddenly did not exist, the universe would be very little changed.

The next level of particles seems to be simply a duplicate of the first, except that the particles are somewhat heavier. All however (except the neutrino) are violently unstable, and the various particles that they make up rapidly disintegrate into level I particles. The third level is just a repeat of the same story.

The question is bound to arise, what are these other levels *for*? Why does nature need them? What role do they play in shaping the universe? Are they merely excess baggage, or do they fit into some mysterious and as yet only dimly-perceived jigsaw? More disturbingly, are there only three levels, or can we expect more — perhaps an unending sequence — to appear in the future as progressively higher energy particle accelerators become available?

Our perplexity is deepened by a further complication. To avoid conflict with a fundamental principle of quantum physics, it is necessary to suppose that each flavour of quark actually comes in three distinct forms, known as 'colours'. Any given quark must be envisaged as a sort of multichrome (figuratively speaking) superposition, continually flashing (again figuratively speaking) from 'red' to 'green' to 'blue'. It all begins to look horribly like a zoo again. But help is at hand. Symmetry comes to the rescue once again, though in a still deeper and subtler form known, appropriately enough, as supersymmetry.

To understand supersymmetry we need to pick up the other strand in this analysis: the forces. Whatever the complexities of the particle zoo, there appear to be only four basic types of force: gravity and electromagnetism, familiar from daily life, and the weak and strong nuclear forces. The strong force between neutrons and protons cannot, of course, be fundamental, because these particles are themselves composites and not elementary. When two protons attract, we really see the combined effect of six quarks interacting. The fundamental force is that between the quarks. It is possible to describe the interquark force after the fashion of the electromagnetic field, with colour playing the role of electric charge. The counterpart of the photon is the so-called gluon, whose job is to glue the quarks together by being continually bounced to and fro between them in the fashion of 'messengers' already described. In analogy to electrodynamics, physicists refer to this 'colour'-generated force-field theory as chromodynamics. Chromatic processes are more complicated than their electromagnetic counterparts for two reasons. First, because there are three colours as against a single type of electric charge; this leads to a total of eight distinct types of gluon compared to a single species of photon. Second, the gluons also carry colour, and thereby strongly interact with each other, whereas photons are uncharged and so oblivious to other photons.

Twenty years ago it occurred to some far-sighted theorists that four fundamental forces of nature seemed too many, and that perhaps they are not all truly independent. After all, in the 1860s Maxwell had produced a mathematical description which unified electricity and magnetism into a single theory of electromagnetism. Perhaps a further synthesis was possible.

Added impetus for this idea came from a class of mathematical headaches that stubbornly refused to go away. Whenever the quantum theory of fields was applied to all but the simplest processes, the answers always turned out to be infinite, and hence meaningless. In the case of the electromagnetic field, a subtle mathematical sleight-of-hand enabled the infinities to be sidestepped, and the theory retains its predictive power for all conceivable electromagnetic processes. But the same trick did not work on the other three forces. One hope was that by combining the electromagnetic force somehow with the other three into a single descriptive scheme, its more mathematically compliant behaviour might rub off on the other forces and enable a sensible formulation to be achieved.

The first step in the realization of this ambitious goal was taken by Steven Weinberg and Abdus Salam in 1967. They succeeded in recasting the mathematical description of the electromagnetic and weak nuclear forces in such a way that the two forces are combined in an integrated mathematical description. Their theory revealed that the reason we normally perceive the electromagnetic and weak forces as distinct (indeed, markedly different in their properties) is because of the extremely low energies employed in our current experiments. Of course 'low' here is relatively speaking: present machines deliver enough punch to a collision that, if it were applied to a billiard ball rather than a proton, the energy release would power the average home for millions of years! Nevertheless, the Weinberg–Salam theory has an inbuilt unit of energy which is only now coming within reach of current technology, and it is gauged against this that the designation 'low' is made.

During the 1970s experimental evidence slowly accumulated in favour of the Weinberg–Salam theory, and in 1980 they received a Nobel prize for their work. In 1971 it was shown that the worrisome infinities were, as hoped, swept aside in the unification, and physicists began to talk of the three, rather than four, fundamental forces of nature.

A major part of this success derives from the appearance in the unified theory of further abstract symmetry groups. It had long been appreciated that Maxwell's beautiful electromagnetic theory owed much of its power and elegance to the balance and symmetry apparent in its mathematical description. Once again the symmetry, known as a *gauge* symmetry, is of an abstract variety, but one reminiscent of daily experience.

Gauge symmetries can be illustrated by the example of the cliff-top ascent. To climb from the bottom to the top of a cliff expends energy. But which strategy is more efficient — to climb the short way, vertically up the face, or take the longer but shallower gradient up the cliff pathway? (see Fig. 24). The answer is that both routes need the same energy (neglecting irrelevant complications of things like friction). In fact, it is easily shown that the energy needed to ascend the cliff is completely independent of the path taken. This is a gauge symmetry.

The example given refers to a gauge symmetry of the gravitational field, for it is the force of gravity that you have to fight to reach the top of the cliff. An identical symmetry applies to electric fields, and something similar, but more complicated, to magnetic fields.

It turns out that the gauge symmetry of the electromagnetic field is intimately related to the massless character of the photon and is the crucial feature in the theory's avoidance of the disastrous infinities. By building a larger gauge symmetry into their unified theory Weinberg and Salam were able to tame the weak force and marry it to electromagnetism.

Spurred on by the success of unified gauge theories, physicists next turned their attention to the other nuclear force — the interquark chromodynamic force. It was not long before a gauge theory of colour was invented, following which attempts were made to unify the

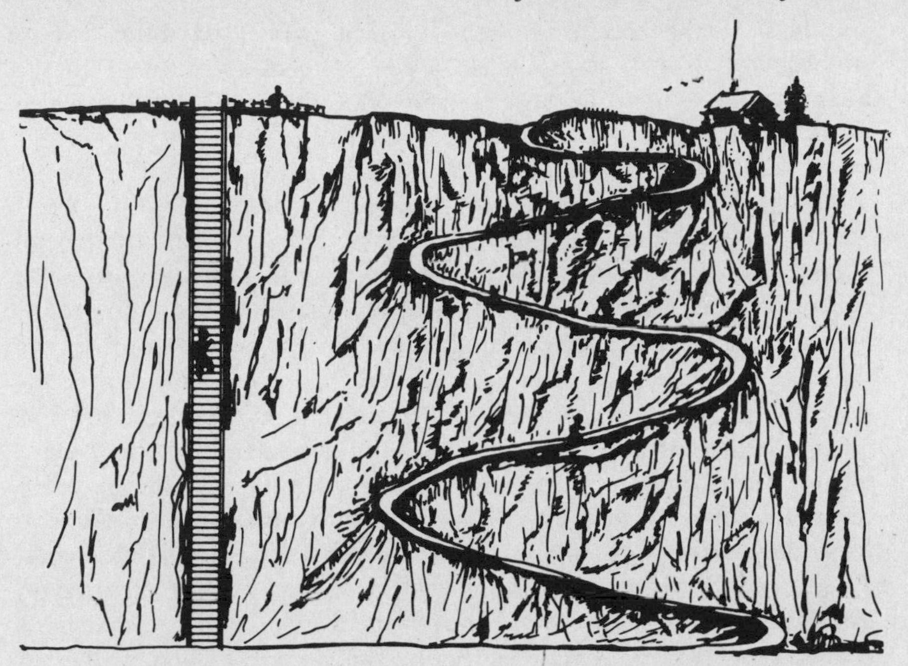

24 A cliff-side ascent illustrates the abstract concept of a 'gauge symmetry'. The total energy expended in reaching the top of the cliff is the same whether the short, hard, vertical route or the long, easy, zig-zag route is selected. This reflects a deep and powerful symmetry of the gravitational field. Similar, but more complex symmetries of nature's other force fields have been exploited in recent mathematical formulations of unified field theories.

electroweak force with the colour force into a 'grand unified theory' (GUT) using an even larger gauge symmetry which rolled all the others into one. As yet, it is too soon to assess the success of GUT, but at least one prediction — that protons might be very weakly unstable and decay spontaneously after an immense duration — is being tested.

This still leaves gravity. The infinity problem plagues gravity with a vengeance, and opinion is shifting more and more towards the view that only in some super-unified theory involving a supersymmetry, will the problem be solved. Attempts to produce such a theory are currently keeping a veritable army of mathematicians and physicists busy. The theory has as its goal the compelling dream of a unified field theory — a single field of force that incorporates within it all the forces of nature: gravity, electromagnetism and the two nuclear forces. But that is only half the story. The fundamental connection between the quantum particles and the forces that act between them implies that any theory of the forces is also a theory of the particles. It follows that a superunified theory should yield a complete description of all the quarks and leptons too, and explain why there are three levels in Table I.

It is sometimes remarked that attaining this dazzling prize would represent the culmination of fundamental physics, for such a theory would be capable of explaining the behaviour and structure of all matter — in a reductionist way, of course. It would enable us to write down all of nature's secrets in a single equation, a sort of master formula for the universe. Such an achievement would confirm the fond belief that the universe runs according to a single, simple, breathtakingly elegant mathematical principle. The compulsion for this ultimate goal has been expressed by John Wheeler in the following terms: 'Some day a door will surely open and expose the glittering central mechanism of the world in its beauty and simplicity.'[1]

So how close are we to achieving this intellectual Nirvana? Theorists are currently pinning their hopes on a set of theories that go under the name of *supergravity*. The pivotal feature of this approach is a bizarre type of supersymmetry that has been cryptically described as the square root of spacetime. What this means is that if two supersymmetry operations are multiplied, you get an ordinary geometrical symmetry operation, such as a sideways shift in space.

At first sight this abstraction might not seem very promising, but closer analysis reveals that supersymmetry is intimately connected with one of the most fundamental attributes that a particle can possess: spin. It is found that all quarks and leptons spin in a rather enigmatic way, the features of which need not concern us here. The point is that the 'messenger' particles — the gluons, photons and the counterparts for gravity and the weak force — either do not spin, or else they do so in a normal rather than an enigmatic way. The significance of supersymmetry is that it connects the enigmatically spinning particles with the others, just as protons were connected with neutrons by isotopic spin symmetry. Thus a supersymmetry operation can change a spinning particle into a non-spinning particle. Of course these 'operations' refer to mathematical procedures. It is not possible to actually change a spinning particle into a spinless one, any more than you can change a left hand into a right hand.

By building a theory of gravity in a supersymmetric framework, the gravitational messenger particle (known as the graviton) acquires companion particles (called gravitinos) which possess the 'funny' sort of spin, and other particles too. The way in which this multiplicity of particles enters into the theory strongly indicates that the dreaded infinities are suppressed, and so far all concrete calculations performed with the theory have yielded finite results.

In the most promising version of supergravity, the entire super-family of particles totals no less than seventy. Many of the particles contained in the theory can be identified with known particles in the real world. Others correspond to particles which may exist but have not yet been discovered. Opinion is still divided, however, about whether there may actually be more particles, hitherto assumed to be fundamental, than this particular theory can accommodate. Some theorists argue that there are simply too many quarks and the time has come to probe deeper and see whether these particles are in turn built out of still smaller units. One argument against the existence of a lower level of structure is that the quarks already inhabit a world some fifteen powers of ten smaller than the atomic nucleus, and this is not far from the ultimate size at which space itself loses meaning. Theory suggests that quantum effects of gravity cause spacetime to break up into a foam on a length scale about twenty powers of ten smaller than a nucleus, and at this point any talk of things 'inside' other things becomes meaningless. So work continues.

It is hoped that this sketchy survey of the elaborate and mind-stretching work currently being undertaken to expose the ultimate structure of matter will at least convey the flavour of research in modern physics. The physicist approaches his subject with something near to reverence, compelled by the belief in the mathematical beauty and simplicity of nature, and convinced that by digging deeper into the bowels of matter, unity will emerge. All experience to date has indicated that the smaller the system probed, the broader are the principles discovered. According to this philosophy, much of the complexity of the world apparent on casual inspection is purely the result of our sampling physical systems at relatively low energy. The belief is that as the energy of the processes of interest soars, so the unity and simplicity become more and more apparent. That is why so much money and effort is devoted to building ultra-high energy particle accelerators — to smash our way into this regime of simplicity.

There was, however, an epoch when this distinctive regime was explored naturally. This occurred during the first split second when the universe exploded into existence in the big bang. At that time, temperatures in excess of a billion billion billion degrees prevailed, equivalent to the colossal energy needed to probe the regime of simplicity. This period is known to physicists as the GUTs era, for its physics was dominated by the processes encompassed by the grand unified theory of the fundamental forces. It was then that the crucial imbalance mentioned in Chapter 3 — the imbalance which led to a tiny excess of matter over antimatter — was established. Then, as the universe cooled, so the unified force separated into the three distinctive forces — electromagnetism, weak and strong — that we perceive now in our relatively cold universe.

The idea that the complex physics of today 'froze' out of the simple physics of the primeval fire has a persuasive elegance. The ultimate principle of nature — the 'glittering central mechanism' sought by Wheeler — is hidden from us for poverty of energy. If one extends these ideas to the epochs that preceded the GUTs era, to still earlier micro-subdivisions of time and yet higher temperatures, the supergravity regime is reached. This represents the very threshold of existence, where space and time themselves become intermingled with the fundamental forces. Most physicists believe that the spacetime concept cannot be continued back within the supergravity era. Indeed, there is even a hint that space and time are to be treated as fields which themselves simply 'froze' out of a primeval soup of pregeometrical elements. Thus, within this supreme era, all four forces of nature would have been indistinguishable, and spacetime would not have yet jelled into a coherent form. The universe would have been merely a collection of ultra-simple components — the raw materials from which God fashioned all of space, time and matter.

The recent developments in the physics of fundamental forces described in this chapter have generated a whole new perspective of nature which is rapidly gaining ground among physicists and astronomers. The universe is coming to be perceived as complexity frozen out of simplicity, in much the same way as the featureless simplicity of the ocean freezes into a tangled ice-flow. There is a feeling among the scientific community that the subjects of cosmology on the one hand and the fundamental forces within matter on the other are coming together to provide a unified description of the cosmos. It is a description in which the ultra-microscopic structure of matter is intimately connected to the global structure of the universe, each influencing the development of the other in a delicate and complex fashion.

The catalogue of successes described in this chapter undoubtedly constitute a triumph for the ideas of modern physics based on reductionist reasoning. By attempting to reduce matter to its ultimate building blocks — leptons, quarks and messengers — physicists have begun to glimpse the fundamental law which controls all the forces that shape the structure and behaviour of matter, and thereby accounts for many of the basic features of the universe.

In spite of this, such an approach to some perceived final truth can only be half the story. We have seen in previous chapters that reductionism fails to account for many observed phenomena that are of a collective or holistic character. It would be ludicrous to try to understand consciousness, or a living cell, or even an inanimate system such as a tornado, in terms of quarks, for example.

Much of the language used in this chapter so far conceals the rather obscure concept which the physicist has concerning structure. When a physicist says a proton is 'made up of' quarks he does not mean it in the literal sense. For example, when we say an animal is made up of cells or a library is made up of books we mean that you can pluck a cell or a book or whatever from the larger system and examine it in isolation. Not so with quarks. As far as we can tell, it is impossible to actually rend a proton asunder and pull out the quarks.

Now there is a respectable history for rendering asunder. Atoms are broken apart literally and routinely; atomic nuclei are harder to smash, but they do disintegrate under high energy impact. This might suggest that firing high-speed projectiles at protons or neutrons will smash them into their constituent quarks. What happens however, is quite different. A high-speed, point-like electron will plough right through the interior of a proton and bounce violently off one of the quarks within, thereby assuring us that there really are quarks inside there somewhere. But if the proton is hit by a sledgehammer rather than a bullet — namely by impact with another proton — instead of seeing quarks spill out among the debris, only more hadrons (protons, mesons and so on) are seen. In other words, quarks never come out in isolation. All that nature seems to permit are collections (twos and threes) of quarks, bound always together.

So when the physicist says a proton is made up of quarks he does not mean that these enigmatic constituents can be individually exhibited. Rather, he is referring merely to a *level of description* which is somehow more fundamental than that of the proton. The mathematical laws that govern the quarks are simpler and more basic than the proton laws. There *is* a sense in which the proton is composite rather than elementary though it is not the same sense in which a library is composite.

Even more severe difficulties emerge when account is taken of the quantum factor for, as we saw in Chapter 8, none of the subatomic particles, quarks or otherwise, are really particles in the common meaning of the word. Indeed, they may not even be 'things' at all.

Once again, the description of matter in terms of such-and-such a collection of particles must really be regarded as levels of description, stiffened by mathematics. The physicist's precise description of the structure of matter is only ever through abstract advanced mathematics, and only in that context can one be precise about the meaning of the reductionist statement 'made up of'.

The difficulties that the quantum factor injects into the subject of 'what is made up of what' is well illustrated by an aspect of the Heisenberg uncertainty principle. This time the dichotomy is not between wave and particle or position and motion, but between energy and time. These two concepts pair together in that mysterious antagonistic way: know one and you un-know the other. If a system, therefore, is inspected for but a brief duration, its energy is likely to fluctuate wildly. In the everyday world, energy is always unalterably fixed; the law of energy conservation is a cornerstone of classical physics. But in the quantum microworld, energy can appear and disappear out of nowhere in a spontaneous and unpredictable fashion.

Quantum energy surges are translated into complex structures when account is taken of Einstein's famous $E = mc^2$ formula. This states that energy and mass are equivalent, or that energy can create matter, a fact already discussed in previous chapters. There the energy was supplied by external sources. Here we wish to discuss material particles being created out of quantum energy fluctuations, without any external input. The Heisenberg principle operates rather like a power bank. Energy can be borrowed for a short duration, provided it is paid back promptly. The shorter the duration the bigger the permitted loan.

In the microworld, a sudden energy surge can cause the fleeting appearance of, say, an electron–positron pair. Its temporary existence is financed by the Heisenberg loan. It never lasts more than a thousand-billion-billionth of a second, but the cumulative effect of countless such 'ghost' particles endows empty space with a sort of shifting texture, albeit a nebulous and insubstantial one. And it is amid this sea of ceaseless activity that subatomic particles must swim. Not only electrons and positrons, but protons and antiprotons, neutrons and antineutrons, mesons and antimesons — all the conceivable particles of nature — partake of this mêlée.

From the quantum angle, an electron is not simply an electron. Shifting energy patterns shimmer around it, financing the unpredictable appearance of photons, protons, mesons, even other electrons. In short, all the paraphenalia of the subatomic world latches on to an electron like an intangible, evanescent cloak, a shroud of ghostly bees swarming around the central hive. When two electrons approach, their shrouds entangle and interaction occurs. The shrouds are just a quantum expression of what were formerly regarded as force fields.

The electron can never be isolated from its retinue of ghost particles. When we ask 'What is an electron?' the answer is not the central particle alone; we must buy the whole package, including the attendant ghost particles which produce the forces. And when it comes to hadrons, which have an internal structure, the identity of the particle becomes even more blurred. A proton somehow contains quarks, which are themselves bound by gluons. There is a sort of Strange Loop here too: forces are produced by particles which in turn produce forces. . . .

In the case of a particle such as a photon, this loopiness means that the photon can display many different faces. By borrowing energy it can temporarily turn into an electron–positron pair, or a proton–antiproton pair. Experiments have even been performed to catch them in the act. Once again, a 'pure' photon can never be distilled out of this complex network of transmutations.

For the vast majority of particles, which are unstable and only live for a minute fraction of a second anyway, the distinction between 'real' and 'ghost' becomes blurred. Who is to say that, for example, a so-called ψ particle, which decays in a thousand-billion-billionth of a second, is real, while an electron–positron pair paid for out of Heisenberg finance with a comparable lifetime, is only a ghost?

Some years ago an American physicist, Geoffrey Chew, likened this restless dance of shadowy transmutation to a democracy. We cannot pin down a particle and say that it is such-and-such an entity. Instead we must regard every particle as somehow made up of every other particle in an endless Strange Loop. No particle is more elementary than any other. (This is the 'bootstrapping' idea mentioned briefly in Chapter 4.)

It will be evident that there is a strong holistic flavour to the quantum aspects of the nature of matter: interlocking levels of description with everything somehow made up of everything else and yet still displaying a hierarchy of structure. It is within this all-embracing wholeness that physicists pursue the quest for the ultimate constituents of matter and the ultimate, unified, force.

12. Accident or design?

> 'Whence arises all that order and beauty we see in the world?'
>
> Isaac Newton

> 'Man at last knows that he is alone in the unfeeling immensity of the universe . . . Neither his destiny nor his duty have been written down.'
>
> Jacques Monod in *Chance and Necessity*

In his book *Natural Theology* William Paley (1743–1805) articulated one of the most powerful arguments for the existence of God:

> In crossing a heath, suppose I pitched my foot against a *stone*, and were asked how the stone came to be there: I might possibly answer, that, for anything I knew to the contrary, it had lain there for ever; nor would it, perhaps, be very easy to show the absurdity of this answer. But suppose I found a *watch* upon the ground, and it should be inquired how the watch happened to be in that place. I should hardly think of the answer I had given before — that, for anything I knew, the watch might always have been there. Yet why should not this answer serve for the watch as well as for the stone?[1]

The intricate and delicate organization of a watch, with its components dovetailing accurately, is overwhelming evidence for design. Someone who had never seen a watch before would conclude that this mechanism was devised by an intelligent person for a purpose.

Paley went on to argue that the universe resembles a watch in its organization and complexity — though on a vastly greater scale.

Surely, therefore, there must exist a cosmic designer who has arranged the world this way for a purpose: 'the contrivances of nature surpass the contrivances of art, in the complexity, subtlety and curiosity of the mechanism'.

The argument from design came to be associated with the concept of *teleology*: the idea that the universe has been programmed to evolve towards some final goal. In its broadest form the teleological argument encompassed both the order of simplicity and the order of complexity. It is an old idea. Aquinas wrote that 'an orderedness of actions to an end is observed in all bodies obeying natural laws, even when they lack awareness . . . which shows that they truly tend to a goal, and do not merely hit it by accident'. Though Aquinas knew nothing about the mathematical simplicity of the fundamental laws of physics, he spotted the striking fact of the compliance of material bodies with orderly laws and used that fact as evidence for a designer God.

The teleological argument was so savagely attacked that it is treated today with circumspection by theologians. Nevertheless it does have some modern proponents. 'The existence of order in the universe', writes Swinburne, 'increases significantly the probability that there is a God.'[2] But Swinburne bases his argument on the order of simplicity rather than the order of complexity. The idea that complex natural structures provide evidence for a cosmic designer seems to have fallen into disrepute.

The main objection to the argument from design involving complexity is that many systems which display complex order and structure can, in fact, be explained as the end result of perfectly ordinary natural processes. This does not, of course, prove that *all* ordered systems have arisen naturally, but it makes us cautious about inferring the existence of a designer purely on the rather superficial grounds that something looks too complicated to have arisen by chance. One must also have some understanding of the processes whereby complex order can develop.

The classic conflict between these opposing philosophies came with Charles Darwin's publication of *The Origin of the Species*. The exquisite organization of living creatures seems to offer the best possible demonstration of a supernatural designer, yet the evidence of biology and geology provide an adequate explanation for the extraordinary characteristics of biological organisms. Evolution of biological order by mutation and natural selection is now accepted virtually unanimously by scientists and theologians alike. Though Darwin's original theory is by no means established in its entirety, the basic principles and mechanisms of evolution are no longer seriously in doubt.

The essential feature of Darwinian evolution is its accidental nature. Mutations occur by blind chance, and as a result of these purely random alterations in the characteristics of the organisms nature is provided with a wide range of options from which to select on the basis of suitability and advantage. In this way, complex organized structures can arise from the accumulation of vast numbers of small accidents. The corresponding increase in order (fall in entropy) occasioned by this trend is more than paid for by the much greater number of damaging mutations which are weeded out by natural selection. There is thus no conflict with the second law of thermodynamics. Today's beautifully fashioned creatures sit atop a family tree festooned with genetic disasters.

Whether one is prepared to accept that the Darwinian mechanism of evolution is the whole story, it cannot be denied that mutation and natural selection must be a major contributory factor in the development of biological order. The essential principle that physical systems can spontaneously organize themselves into intricate complexity is an empirical fact. In Chapter 5 we saw how many simpler examples of self-organization in the laboratory have been studied by physicists and chemists in recent years. Indeed, so important have these studies become that a new word — synergetics — has been coined to describe them. The conclusion must be that the presence of order in a system, however remarkable and complex it might be, is in itself no guarantee that a designer is necessary. Order can, and does, occur spontaneously.

These observations still, however, leave open a vital issue. Though the spontaneous appearance of order will not conflict with the second law of thermodynamics so long as compensatory disorder is generated elsewhere, it is clear that no order at all could exist unless the universe as a whole started out with a considerable stock of negative entropy. If total disorder always increases, in accordance with the second law, then the universe must, it seems, have been created in an orderly condition. Does this not provide strong evidence in favour of a creator-designer? After all, even if natural processes can generate localized order unaided, a fund of negative entropy is still needed to drive those processes in the first place. True, this could only constitute evidence of a designer-by-proxy, a creator who winds up the machine and then lets it crank out whatever structures it will, but even that strategy would involve supernatural dexterity of an astonishing degree, for the following reason.

Entropy, or disorder, is closely related to the concepts of probability and arrangement. A high-entropy, or disordered system, is one that can be achieved in a large variety of ways. Consider, for example, a box of gas in equilibrium at some uniform temperature and density; this is the condition of maximum possible entropy for the gas. Under these circumstances the molecules of gas could be rearranged in a huge variety of ways (for example, by being moved to different positions, or having their velocities altered) without affecting the large-scale properties of the gas. On the other hand, consider a very low entropy state, where all the molecules are moving on parallel trajectories, or another in which every molecule is crowded into one end of the box. These ordered configurations are exceedingly sensitive to any minute rearrangement of the molecules, and can only be achieved by a very limited fraction of the total available number of molecular arrangements. It follows that ordered (low-entropy) states are highly improbable and unstable. They require the careful cooperation of vast numbers of individual molecules. In disordered (high-entropy) states, all the molecules can move about randomly without regard for the others.

Now if you were asked to pick an arrangement of molecules at random, it is overwhelmingly likely that you would choose one that corresponds to maximum entropy, simply because there are vastly more possible disorderly arrangements than orderly ones. It is rather like the monkey who tinkers at random on a piano. The chances of his playing a well-known tune rather than a chaotic sequence of notes is minute. A mathematical investigation shows that order is exponentially sensitive to rearrangements. That is to say, the probability of a random choice leading to an ordered state declines exponentially with the degree of negative entropy. An exponential relation is characterized by its rapid rate of growth (or decline). For example, a population that grows exponentially doubles its size in a fixed interval of time: 1, 2, 4, 8, 16, 32 . . .

The exponential factor implies that the odds against randomly-generated order increase astronomically. For example, the probability of a litre of air rushing spontaneously to one end of a box is of the order $10^{10^{20}}$ to one, where the number $10^{10^{20}}$ stands for one followed by 100,000,000,000,000,000,000 zeros! Such figures indicate the extreme care with which low-entropy states must be selected from the vast array of possible states.

Translated into a cosmological context, the conundrum is this. If the universe is simply an accident, the odds against it containing any appreciable order are ludicrously small. If the big bang was just a random event, then the probability seems *overwhelming* (a colossal understatement) that the emerging cosmic material would be in thermodynamic equilibrium at maximum entropy with zero order. As this was clearly not the case, it appears hard to escape the conclusion that the actual state of the universe has been 'chosen' or selected somehow from the huge number of available states, all but an infinitesimal fraction of which are totally disordered. And if such an exceedingly improbable initial state was selected, there surely had to be a *selector* or *designer* to 'choose' it?

A useful image here is that of a creator equipped with a pin. Before him is a vast 'shopping list' of universes, each characterized by their initial state. If the creator picks a universe by sticking in a pin at random, there is an overwhelming probability that the choice will be a highly disordered cosmos with no appreciable structure or organization. Indeed, to find an ordered universe, the creator would have to scour a selection of 'models' that is so vast its number could not be written down on a sheet of paper as big as the entire observable universe.

The mystery of how the universe got into its low-entropy state has exercised the imagination of several generations of physicists and cosmologists, many of whom have been reluctant to appeal to divine selection. The pioneer of statistical thermodynamics, Ludwig Boltzmann, preferred to fall back on blind chance. He suggested that the cosmic order had arisen spontaneously as a result of a collaboration of unbelievably rare fluctuations from equilibrium. The basis of his argument is the fact that, even in equilibrium, the molecules of a gas do not remain inert, but are continually rushing about in a random sort of way. From time to time, purely by chance, a few molecules will find themselves in unwitting cooperation, and a tiny enclave of order will arise, fleetingly, amid an ocean of chaos. Exponentiate the time scale and one can be persuaded that still larger regions of cooperation will eventually accidently occur. If the universe has enough time available then, sooner or later, one might suppose, whole stars, whole galaxies, will simply happen to form — accidentally. The fact that the time for such an absurdly improbable accident is inconceivably long (at least $10^{10^{80}}$ years) is, in principle, no problem if one is prepared to believe that the universe is of infinite age.

According to this view, the universe spends the overwhelming majority of its time in total chaos, with no organization whatever. But from time to time, after intervals of mind-numbing duration, there occurs a few billion years of accidental order. The reason that we —

humanity — are present to witness one such occurrence of staggering improbability is simply because, in the absence of such a 'miracle', life could not exist. Because life feeds on negative entropy (see Chapter 5) conscious observers will only exist in the epochs of 'miraculous' fluctuations.

An interesting by-product of Boltzmann's reasoning is its assurance of a form of immortality. The continual molecular reshuffling which is responsible for the universe 'winding itself up' can be proved, mathematically, to possess the following curious property. As the molecules mill around, the universe visits state after state. Eventually, every possible state that can ever exist will be visited: anything at all that can happen, will happen, sooner or later. After this, the shuffling continues, and the universe starts *revisiting* states that have previously occurred. Eventually, every state will have been revisited, and so the process continues. This phenomenon of unlimited repetition and duplication is known as the Poincaré cycle, after the mathematical physicist Henri Poncaré who proved the result (at least, he proved it for an idealized model). Taken at face value, the Poincaré theorem implies that, in the fullness of time, planet Earth, long since disappeared, will be reconstituted, together with all its inhabitants! Moreover, this will happen infinitely often. But for every instance of more or less exact duplication there will be untold numbers of cases in which there are departures from the present arrangement. The closer the 'fit' the smaller the probability of reconstitution and the longer the wait.

Few physicists would take Boltzmann's explanation of cosmic order seriously. The basic mechanism of Poincaré recycling is not in doubt, however it is now known that the universe is not simply sitting there, shuffling away, but is in a state of global expansion. It is generally assumed that this expansion forces the universe to be of finite age. Its multi-billion-year lifetime is but an insignificant drop in the ocean of time needed to bring about all but a trifling decrease in entropy.

Boltzmann's argument does, however, bring out one vital feature of enduring value. The universe we perceive is, necessarily, selected *by us*, from the elementary requirement that life, and hence consciousness, can only develop under the appropriate physical conditions. By definition, we cannot observe an uninhabitable cosmos. This simple fact has, as we shall shortly see, been used to argue by some that the extraordinarily improbable, low-entropy universe that we observe has indeed been selected from a vast array of possible universes (nearly all of which are disordered); but the selection has been made by *us*, not by God.

Adopting, therefore, the big bang scenario, it would seem that we have little choice but to assume that the universe went bang in a remarkably orderly way, even though an accidental creation would, with a probability that is a virtual certainty, produce a totally disordered universe. This fundamental paradox of cosmology has stimulated several different responses:

1. *So what?*

Many scientists incline to the view that it is meaningless to discuss the concepts of probability, randomness and likelihood on an *a posteriori* basis. If you pick up a pebble on a beach at random, and carefully measure its size and shape you could correctly conclude it was wildly improbable that you had selected a pebble of those exact dimensions. But you would not be justified in proceeding to claim it must have therefore been a miracle that you made the choice you did, or that some supernatural or occult agency had been responsible for guiding your selection. Such arguments carry no conviction *after the event*. Amazement would have been justified had the pebble's dimensions been specified in advance, of course. In the same vein it could be argued that, given the existence of the universe, its particular structure need occasion no surprise: it simply is the way it is.

A related difficulty is that, at least according to one conception, probability is only defined in relation to a collection of trials. To say that, for example, the toss of a die will produce the result 'two' with a probability one-sixth is to say that, after many, many such tosses, roughly one-sixth of them will have produced a 'two'. The greater the number of trials the closer the proportion converges on the value one-sixth. At the very least, the subject to which our discussion of probability is directed must be a member of a collection or ensemble of similar things. The face of a die, for example, has five neighbours; the pebble on the beach had millions of neighbours. If, then, there is but one universe, what meaning can be attached to discussions of its likelihood?

This argument is not completely convincing though. If the selected pebble had turned out, for example, to be exactly spherical, surprise would indeed have been justified, even if its spherical nature had not been specified in advance. A sphere is a very special sort of shape with the property that it is mathematically highly regular. Even after the event the random selection of an exactly spherical pebble would be regarded as a remarkable circumstance deserving some sort of explanation. Likewise, a universe that is suitable for human habitation has a special significance for us that is absent for the vast majority of other possible universes: those that are uninhabitable.

At this point the 'so what' proponents reply that, had the universe not been arranged in the way it is, we should not be here to marvel at it. Indeed, any universe in which intelligent creatures can frame philosophical and mathematical questions is, by definition, a universe of the sort we observe, however remarkable that universe would otherwise have been *a priori*. In other words, they maintain, there is not, after all, anything very extraordinary or mysterious about the highly ordered universe we perceive, because we could not (obviously) perceive it otherwise.

This type of reasoning receives some support from the philosophy of logical positivism which argues, crudely speaking, that it is meaningless to talk about what can never be observed. What does it mean to discuss a universe that has no conscious observers within it? As such a universe could never be verified or refuted by observation, its existence would seem to have no meaning or significance for conscious individuals.

A related argument is that of the so-called strong anthropic principle, first articulated in detail by the astrophysicist Brandon Carter, and much discussed in recent years by physicists and astronomers. According to this principle 'The universe *must* be such as to admit conscious beings in it at some stage'[3] (my italics). This amounts to saying that, far from being astonishingly unlikely, the universe had *no choice* about appearing with the appropriate degree of order required for life to appear.

These two positions — logical positivism and the strong anthropic principle — hinge on the paramountcy of human (or extraterrestrial) intelligent observers. The theologian could counter, however, that God is an observer, who moreover does not require special physical conditions for his existence. So universes that never produce life are still meaningful if observed by God.

2. *The many-universes theory.*

According to this point of view, there *is* an ensemble of universes of which ours is but one member. The universe we perceive is only one of a huge, perhaps infinite, collection of universes, each differing from the other members of the ensemble in some way. Somewhere among this collection is an example of every possible arrangement of matter and energy. Although the overwhelming majority of these universes are unsuitable for life, and lie very close to the totally chaotic conditions of maximum entropy (thermodynamic equilibrium), nevertheless there will be a minute fraction in which, accidentally, the conditions came out just right, and life develops. Obviously it is only those accidental universes that will be perceived by living organisms, who will write books about how incredibly improbable their world is.

Boltzmann's hypothesis, mentioned above, is logically identical to the many-universes theory. His universes occur sequentially, but the organized phases are separated by such enormous chasms of time that they are all but physically independent. A modern variant of the sequential scenario is the oscillating universe theory. As we shall see (in Chapter 15) the present expansion of the universe might not continue indefinitely. If it does not, the universe will eventually begin to contract, falling back on itself in a gigantic cataclysm known as the 'big crunch'. Some physicists have speculated that the highly compressed cosmos, rather than imploding to oblivion at a spacetime singularity, will 'bounce' at some enormous density, and thereafter embark on a new cycle of expansion and eventual contraction. In this scenario, then, the universe continues indefinitely in a cyclical fashion, oscillating between collapsed 'crunch-bangs' and distended, low-density states, like a balloon being successively inflated and deflated.

The oscillating universe suffers from the physical problems associated with all infinitely old universes which were discussed briefly in Chapter 2. However, the uncertainties surrounding the physics of the extremely collapsed state widen the scope for speculation, and one suggestion, due to Wheeler, is that the 'crunch-bangs' have the effect of 'reprocessing' the cosmos. What this means is that each new cycle of expansion and contraction is a sort of 'new deal' in which the physical conditions are re-scrambled randomly. No attempt is made to explain how this might happen, but if it did it would clearly enable the universe to explore all possibilities open to it after a sufficient number of cycles — which would, of course, have to be astronomically large. Once again, only in those cycles where, by accident, the cosmos scrambler got it just right, would cosmologists evolve to speculate about it.

An alternative to assuming an ensemble of universes in time is to suppose that there is only one universe, which is infinite in spatial extent. Almost all of the cosmos would be close to equilibrium (no structure or organization) but, here and there, oases of order would appear spontaneously out of the chaos, by chance fluctuations. The distances between the oases would be inconceivably great, of course, but life and conscious observers could only form within an oasis, so all observers of this universe would necessarily perceive order.

Perhaps the most popular version of the many-universes idea

comes, however, from Everett's interpretation of the quantum theory. In this theory *all possible* quantum worlds are actually realized, and coexist in parallel with each other. Thus, every time an electron faces two choices *both* alternatives occur, and the entire universe divides into two. Each universe is complete with inhabitants (whose brains and presumably minds have also bifurcated) each set of which believes that the electron has abruptly opted for one of the alternatives. The two universes are disconnected from each other in the sense that it is not possible to travel from one to the other through ordinary space or time. They exist 'side-by-side' or 'in parallel' in some abstract sense. And because there are as many universes as there are quantum choices, every possible arrangement of matter and energy will occur somewhere among the infinite array of parallel worlds.

This pattern of reasoning — that observers select a highly atypical universe from among a vast number of alternatives — is known as the weak anthropic principle. The idea has been attacked on a number of philosophical and physical grounds. First it is, in a sense, too successful. By allowing nature to realize all possibilities, anything at all might be 'explained'. Indeed, we might need no science at all. It is merely necessary to make a case that such-and-such a feature is indispensible to human existence and, hey presto, it is explained.

Another weakness of the anthropic argument is that it seems the very antithesis of Occam's razor, according to which the most plausible of a possible set of explanations is that which contains the simplest ideas and least number of assumptions. To invoke an infinity of other universes just to explain one is surely carrying excess baggage to cosmic extremes, not to mention the fact that all but a minute proportion of these other universes go unobserved (except by God perhaps). 'Not a bit of it,' counter the anthropic proponents. 'The Everett interpretation of the quantum theory may be expensive in universes but it is extremely cheap in epistemology. Consider the convoluted and implausible assumptions that are made in the alternative explanations of the quantum measurement problem. In the many-universes theory, the interpretation just drops out of the formalism with no additional metaphysical hypotheses.'

Nevertheless, the many-universe theorists concede that the 'other worlds' of their theory can never, even in principle, be inspected. Travel between quantum 'branches' is forbidden. Moreover, the ordered regions in the infinite or oscillating model universes are separated by such huge expanses of space or time that no observer can ever verify or refute empirically the existence of the many universes. It is hard to see how such a purely theoretical construct can ever be used as an *explanation*, in the scientific sense, of a feature of nature. Of course, one might find it easier to believe in an infinite array of universes than in an infinite Deity, but such a belief must rest on faith rather than observation.

The scientific basis of both the weak and strong anthropic principles has also been challenged. The appeal to the concept of probability, upon which the whole anthropic argument turns, has been used to argue against it. The issue concerns the relative likelihood of small versus large fluctuations. Imagine the chimpanzee on the piano again, tinkering randomly. After an exceptional wait we might reasonably expect to hear a three or four note sequence of a familiar tune. The wait for, say, a six note sequence would be immensely longer. The improbability rises sharply with the degree of order involved. To take another example, a shuffled pack of cards might well result in each of four players receiving an ace. Less likely is for each to receive an ace, two and three of the same suit. The odds against each player being dealt a whole suit are colossal. Small coincidences are relatively much more likely than big ones.

In the context of cosmology, a random accident which produces, say, one star, is exceedingly more probable (less improbable) than one which produces a whole galaxy. And the chances of billions of galaxies forming this way would be infinitesimal compared to the chances for a single galaxy. But, it has been reasoned, surely only one galaxy — perhaps only one star — would be sufficient for life to form and observers to arise? Why, then, do we observe an entire universe filled with orderly structure? In the many-universes theory for example, there would be untold billions of universes with just one galaxy for every universe which had two; when more galaxies are involved the proportional discrepancy escalates rapidly. If observers exist in all these universes, the overwhelming majority will, therefore, inhabit a one-galaxy as opposed to a many-galaxy cosmos. How, then, do we account for the existence of so many galaxies in *our* universe?

The only conceivable answer to his criticism is that, for some as yet unknown reason, the formation of a galaxy is somehow linked to the large scale structure of the universe. Perhaps galaxies can only form when some special global condition is realized, and then when this is the case, they form everywhere. In other words, universes either have galaxies everywhere, or no galaxies. Linking principles of this type are known in physics but the mechanism of galaxy formation is still too obscure for a realistic evaluation of such a possibility.

3. *Order out of chaos.*
The third response to the mystery of the origin of cosmic order is an attempt to demonstrate that it has somehow arisen out of an initially chaotic state as a result of natural physical processes (not merely inconceivably rare fluctuations). (This idea has already been discussed in detail in Chapter 4; only a brief summary will be given here.) At first sight such an approach seems doomed to failure. Does the second law of thermodynamics not state that (fluctuations aside) order can give way to chaos but not vice versa?

It does indeed, but one has to look at the small print. Strictly stated, the second law is intended to apply only to completely isolated systems. Obviously any portion of the universe, however large, is not isolated, because it is in contact with the surrounding portions. More important, the entire universe is subject to the famous expansion, and this external disturbance can make all the difference.

A good analogy here is the humble piston and cylinder of the type found in ordinary petrol-driven engines. Imagine a gas confined in the cylinder beneath the piston. If the piston is at rest, the gas will be in equilibrium at a uniform temperature and pressure — a condition of maximum entropy. No further change can be expected: the gas is devoid of any ordered structure or organized activity. Suppose now that the piston is abruptly raised, allowing the gas to expand. Suddenly the gas is no longer uniform. The density is lower near the retreating piston where more available space is opening up. Turbulent motions occur as the gas flows forward into this space. If the piston should then reverse and return to its starting position the gas would eventually settle down again into a new state of thermodynamic equilibrium, but the entropy will have risen as a consequence of this disturbance. Temporarily the gas will have grown a structure and organization as the piston moved.

Have we found a loophole in the second law? No. The entropy of the gas still rises after a complete cycle of motion (it is hotter). The initial state of equilibrium *was* the maximum entropy state consistent with the external constraints on the system. When the piston moved, however, those constraints changed, allowing the gas to seek a still higher entropy state. In short, the initial state of equilibrium was only a relative, not an absolute maximum.

In the cosmological case the expansion of the universe plays a similar role to the piston as a changing external constraint. Cosmologists point out that, far from being in an orderly state, the primeval universe was close to thermodynamic equilibrium. None of the familiar structures we now observe — galaxies, stars, atoms — were present in the big bang. Indeed, before about one minute or so after the beginning, the temperature was too hot even for atomic nuclei to exist. Somehow the present orderly structure has arisen from the primeval chaos. How?

Most of the complex organization with which we are familiar on Earth, such as biosystems and weather patterns, are generated by sunlight, the vital source of negative entropy on which we all feed. The sun's store of negative entropy is its nuclear fuel (mainly hydrogen). The most relaxed, high-entropy form of nuclear matter consists of medium-mass elements, such as iron. The production of sunlight represents the entropy produced by the sun's attempt to convert hydrogen into iron through a succession of nuclear reactions. The secret of the sun's order (negative entropy), and that of most other stars, is to be found in the explanation for its hydrogen content. About three-quarters of the mass of the universe is made of hydrogen, nearly all the rest consisting of the next lightest element, helium. Why is it not all made of iron?

The answer to this question was given in Chapter 4. The primeval universe was simply too hot for iron to exist, and its subsequent cooling was too rapid to allow significant nuclear transmutation to take place. The primeval material thus remained trapped in the form of low-entropy hydrogen, unable to achieve its goal of high-entropy iron until the stars appeared.

By appealing to an explanation along these lines, it is evidently unnecessary to suppose that the universe was created in a remarkably ordered state after all. The primeval material was actually in a condition of total disorder (maximum entropy). Such a state can be realized in the greatest number of ways, and the creator-with-the-pin would merely need to stab the 'shopping list' at random. The mystery of the origin of the cosmic order is solved.

Or is it?

The nuclear condition of the cosmic material is certainly a crucial factor in generating the observed structure and organization, but it is not the whole story. The larger structures — stars and galaxies — are shaped by gravity. Moreover, the crucial cosmic expansion is also controlled by gravity. What can be said about the gravitational organization of the cosmos? Do we live in a highly ordered, or a disordered universe, from the gravitational point of view? These questions will form the subject of the next chapter.

13. Black holes and cosmic chaos

'Chaos is ubiquitous.'

John Barrow

Was our universe created in a very special state, carefully fashioned so that, in the fullness of time, life and eventually mind would blossom forth to marvel at it? Or do we live amid a monstrous and meaningless accident, a cosmic eruption from nothing, that has occurred purely at random? Surely there can be no more pressing task for today's cosmologist than to tackle that central question of existence.

In the previous chapter arguments were presented which indicate that, despite the legislative imperative of the second law of thermodynamics, much of the cosmic order could have arisen naturally and unremarkably from a primeval universe which was totally chaotic, and completely consistent with a random and accidental origin of the physical world. When account is taken of gravity, however, the picture changes sharply.

Gravity is the weakest of nature's forces, but being cumulative in power it dominates on the large scale. We look to gravity to explain the structure of star clusters and galaxies, as well as the global motion of the expanding cosmos. Although the nature of gravity is well understood in terms of Einstein's general theory of relativity, with its spacewarps and timewarps, when it comes to the concept of gravitational *order*, physics is floundering. There is still no agreement or understanding of the thermodynamics of gravitating systems, and concepts such as the entropy of a gravitational field are still only vaguely formulated.

As explained in Chapter 4, a paradoxical aspect of gravitational entropy is that, what appears to us as a more structured state is actually at a higher entropy than a less structured state. For example, an initially uniform distribution of stars will relax into a more complicated organization, with a high density of rapidly moving stars lying close to the centre of gravity, and a more diffuse population of slower stars surrounding it (see Fig. 7). This tendency for gravitating systems to grow structure spontaneously is a good example of self-organization. It should be contrasted with the behaviour of a gas in which gravitational forces are negligible. The gas tends towards a state of uniformity, with a homogeneous temperature and density throughout. Gravitating systems, however, became clumpy and inhomogeneous.

In the absence of other forces, all gravitating systems would completely collapse. The Earth, for example, is only held up against its own weight by the stiffness of its material (ultimately electrical in origin). Likewise the sun avoids implosion only because of the huge central pressure generated by the nuclear furnace in the core. Remove these internal forces, and both these bodies would shrink in minutes, at an accelerating rate. As they did so, their gravity would climb and the rate of shrinkage would accelerate. Soon they would be engulfed in an escalating timewarp and turn into black holes. From the outside, time would appear to stand still, and no further change would be discerned. The black hole represents the equilibrium end state of a gravitating system, corresponding to maximum entropy.

Although the entropy of a general gravitating system is not known, work by Jacob Bekenstein and Stephen Hawking, in which the quantum theory is applied to black holes, has yielded a formula for the entropy of these objects. As expected, it is enormously greater than the entropy of, for instance, a star of the same mass. Assuming that the relationship between entropy and probability extends to the gravitating case, this result may be expressed in an interesting way. Given a random distribution of (gravitating) matter, it is overwhelmingly more probable that it will form a black hole than a star or a cloud of dispersed gas. These considerations give a new slant, therefore, to the question of whether the universe was created in an ordered or disordered state. If the initial state were chosen at random, it seems exceedingly probable that the big bang would have coughed out black holes rather than dispersed gases. The present arrangement of matter and energy, with matter spread thinly at relatively low density, in the form of stars and gas clouds would, apparently, only result from a very special choice of initial conditions. Roger Penrose has computed the odds against the observed universe appearing by accident, given that a black-hole cosmos is so much more likely on *a priori* grounds. He estimates a figure of $10^{10^{30}}$ to one.[1]

The absence (or at least lack of predominance) of black holes is not the only issue. The large scale structure and motion of the universe is equally remarkable. The accumulated gravity of the universe operates to restrain the expansion, causing it to decelerate with time. In the primeval phase the expansion was much faster than it is today. The universe is thus the product of a competition between the explosive vigour of the big bang, and the force of gravity which tries to pull the pieces back together again. In recent years, astrophysicists have come to realize just how delicately this competition has been balanced. Had the big bang been weaker, the cosmos would have soon fallen back on itself in a big crunch. On the other hand, had it been stronger, the cosmic material would have dispersed so rapidly that galaxies would not have formed. Either way, the observed structure of the universe seems to depend very sensitively on the precise matching of explosive vigour to gravitating power.

Just how sensitively is revealed by calculation. At the so-called Planck time (10^{-43} seconds) (which is the earliest moment at which the concept of space and time has meaning) the matching was accurate to a staggering one part in 10^{60}. That is to say, had the explosion differed in strength at the outset by only one part in 10^{60}, the universe we now perceive would not exist. To give some meaning to these numbers, suppose you wanted to fire a bullet at a one-inch target on the other side of the observable universe, twenty billion light years away. Your aim would have to be accurate to that same part in 10^{60}.

Quite apart from the accuracy of this overall matching, there is the mystery of why the universe is so extraordinarily uniform, both in the distribution of matter, and the rate of expansion. Most explosions are chaotic affairs, and one might expect the big bang to have varied in its degree of vigour from place to place. This was not so. The expansion of the universe in our own cosmic neighbourhood is indistinguishable in rate from that on the far side of the universe.

This coherence of behaviour over the whole cosmos seems all the more remarkable when account is taken of what are known as light horizons. When light spreads out across the universe it has to chase the retreating galaxies which are being swept apart by the expansion. The rate of recession of a galaxy depends on its distance from the observer. Distant galaxies recede faster. Imagine a flash of light emitted from a particular place at the instant of the creation. The light will have travelled about twenty billion light years across space by now.

Regions of the universe farther away than this will not yet have received the light. Observers there would not be able to see the light source. Conversely, observers near the light source would not be able to see those regions. It follows that no observer in the universe can see beyond twenty billion light years at this time. There is a sort of horizon in space, which conceals everything that lies beyond. And because no signal or influence can travel faster than light, it follows that no physical connection at all can exist between regions of the universe that lie beyond each other's horizon.

When telescopes are turned on the outer limits of the observable universe, they probe regions that have apparently never been in causal contact with each other. The reason is that distant regions which lie on opposite sides of the sky as viewed from Earth are so far apart from each other that they are beyond each other's horizon. The situation is closely analogous to ordinary horizons. A lookout on a ship at sea may just be able to discern two other ships — one ahead, one astern — near his horizon, but these other ships will be invisible from each other because of their greater separation. Similarly, the remote galaxies which lie on opposite sides of the sky are located beyond each other's light horizon. Because all physical influences or communications are limited by the speed of light, it is not possible that these galaxies can have coordinated their behaviour.

The mystery is, why are those regions of the universe that are causally disconnected so similar in structure and behaviour? Why do they contain galaxies of the same average size and form, retreating from each other at the same rate? The mystery becomes all the more profound when we realize that this behaviour is a remnant of long ago when the galaxies first formed. But in the past light had travelled less far since the creation, so the horizons were closer. At one million years they were a million light years across, at one hundred years a hundred light years, and so on. If we go back to the Planck time again, the horizons were a mere 10^{-33} cm in size. Even allowing for the expansion of the universe, regions as small as this would not, according to the standard theory, have swelled to a visible size by now. It seems that the entire observable universe was, at that time, separated into at least 10^{80} causally disconnected regions. How is it possible to explain this cooperation without communication?

A related problem is the extreme degree of cosmic isotropy: uniformity with orientation. Looking outwards from Earth, the universe presents the same aspect on the large scale in whichever direction we choose to look. Careful measurements of the relic cosmic background heat radiation show that the incoming flux is accurately matched from all sides to better than one part in a thousand. Had the big bang been a random event, such exceptional uniformity would be almost impossibly unlikely.

The upshot of these considerations is that the gravitational arrangement of the universe is bafflingly regular and uniform. There seems to be no obvious reason why the universe did not go berserk, expanding in a chaotic and uncoordinated way, producing enormous black holes. Channelling the explosive violence into such a regular and organized pattern of motion seems like a miracle. Is it? Let us examine various responses to this mystery:

1. *Hidden principle*.

When a quantity is found to have a value very close to zero, physicists are inclined to suspect that it is exactly zero for some profound reason. They look for a fundamental principle which will guarantee that the quantity must be precisely zero. For example, there is no discernible difference between the electric charges carried by different electrons. Hence it is concluded that the charges are precisely the same, i.e. their differences are exactly zero. This is a consequence of the fundamental principle of indistinguishability of electrons. Another example is that all objects dropped together hit the ground together (in the absence of air resistance). The difference in their arrival time is taken to be exactly zero, a consequence of what is known as the equivalence principle, a fundamental principle of gravity which requires that the gravitational response of a body is independent of the nature of that body.

One could envisage a principle (or set of principles) which required, for example, the explosive vigour of the big bang to *exactly* match its gravitating power everywhere, so that the receding galaxies just escaped their own gravity. This would imply that the universe expands in such a way as to be precisely on the dividing line between complete dispersal of the cosmic material, and an eventual halt to the expansion, followed by collapse. Such a principle would also ensure that the universe emerged from the big bang with uniformly distributed material, rather than black holes. Likewise, such a principle could ensure that the expansion was exactly uniform in all directions. Although we have no idea what these principles might be, from the fact that the differences in expansion rates in different regions and different directions is very close to zero, it is tempting to suggest that there is a principle of nature which forces these differences to be exactly zero.

Unfortunately, it cannot be that simple. If the universe were *exactly* uniform, then no galaxies would have formed anyway. According to present understanding, it seems that the growth of galaxies from the primeval gases can only have occurred in the time available since the creation if the rudiments of galaxies were present at the outset. The accumulation of material by accretion from the surrounding universe is very slow when it has to compete with the cosmological expansion. Only if the galaxies had a head start would they have overcome the dispersing tendency of the expansion. If a fundamental principle does exist, it seems that it must allow just enough deviation from uniformity to permit the growth of galaxies, but not so much as to produce black holes. A delicate and complicated balancing act indeed!

2. *Dissipation.*

One possible explanation for the uniformity of the cosmic expansion is to suppose that the universe started out with a highly nonuniform motion, but somehow dissipated the turbulence away. Theoretical studies do indeed suggest that a universe which expands much more rapidly in one direction than others is subject to a braking effect due to a variety of mechanisms. For example, the creation of matter from the expansion energy (see Chapter 3) would deplete the vigour of the motion in the rapid direction, and tend to bring it into alignment with the other directions. Other braking processes are also known.

Two objections have been raised against this scenario. The first is that, however efficient the dissipation of primeval turbulence may be, it is always possible to find initial states which are so grossly distorted that a vestige will remain, in spite of the damping. At best one can only succeed in showing that the universe must have belonged to a class of remarkable initial states.

The second objection is that all dissipation generates entropy. The violence of the primeval turbulence would be converted into enormous quantities of heat, far in excess of the observed quantity of the primeval heat radiation. There is, however, a loophole in this objection. The quantity of heat in the universe as such is a meaningless concept. It must be measured against some gauge or standard. The only available comparison is with matter, so cosmologists think in terms of the heat per atom, or more accurately the heat per proton. That is, they compute the total heat in some large volume of space, estimate the mass of matter in that volume and calculate the corresponding number of protons. The heat per proton turns out to be rather small. You would need nearly a million billion times more to rival the heat output of a match. This modest value is, so the objection goes, a consequence of the quiescent nature of the primeval universe. Had it been turbulent, space would now be filled with searing heat radiation. But the loophole concerns the use of protons to gauge the value of the heat. Protons may not be the indestructible particles necessary to provide a fixed gauge for comparison. According to the so-called grand unified theories of the fundamental forces, protons can decay. They can also (by the reverse process) be created. We have seen, in Chapter 3, how protons were created out of the primeval energy, and that the grand unified theories predict (correctly) the heat per proton in terms of the parameters of that theory. Because these theories automatically adjust the proton abundance to fit the available heat, the heat per proton will always work out the same in the end, regardless of how much initial heat is deposited by dissipation of turbulence. Thus the issue of whether or not the universe began in a quiescent, highly uniform state, or one of extreme turbulence and irregularity, hinges on forthcoming verification or otherwise of the grand unified theories, perhaps through the confirmation of proton decay.

3. *Anthropic principle.*

Because a universe full of black holes, or turbulent large scale motions is unlikely to be conducive to life, there is clearly room for an anthropic explanation of the uniformity of the universe. If the weak anthropic principle is to be used, one may envisage an ensemble of universes covering every possible choice of initial expansion motion and distribution of matter. Only in that minute fraction which comes close to the arrangement in the observed universe would life and observers form. Anisotropic or highly inhomogeneous universes would not be cognizable.

To be successful, this explanation would have to demonstrate that even a slight increase in irregularity would be inimical to life. The sensitivity of present physical conditions in the universe to small alterations in the primeval state might be very great. For example, if protons do not decay, even a minute amount of primeval anisotropy could produce so much heat that life would be impossible. Even a hundredfold increase in the cosmic background temperature would spell disaster for life as we know it. However, no detailed calculations have been performed, and the anthropic argument here is open to the same criticisms as those advanced in the previous chapter.

4. *Inflation.*

Very recently an entirely new approach to the cosmic uniformity problem has been suggested. It originates with the grand unified theories, and depends crucially on a number of assumptions about ultra-high energy matter which are debatable, and in any case hard to verify. Nevertheless it vividly demonstrates how an advance in fundamental physics can change our whole perspective of the origin of order in the universe.

It will be recalled that as the universe cooled so the three forces of nature — electromagnetism, weak and strong nuclear forces — 'froze' out of an initially undifferentiated phase into their present distinct form. This phase transition is akin to the change from steam to water or water to ice. The two phases differ not only in the nature of the forces, but also in their gravitational effect. The same mechanism which is responsible for splitting the grand unified force into the separate electromagnetic and nuclear components is also responsible for generating a huge repulsive gravitational force.

The possibility of a sort of cosmic repulsive force was actually invented by Einstein in 1917, although he never really liked the idea and there is no astronomical evidence that one exists today. Nevertheless, the grand unified theories suggest that a cosmic repulsion must inevitably have been present in the hot primeval stage, before about 10^{-35}s, when the universe had the unthinkably high temperature of 10^{28}K (degrees absolute). It has been pointed out by Alan Guth of the Massachusetts Institute of Technology that the existence of this force would have had a dramatic and profound effect on the structure of the primeval universe.

As the universe expanded and cooled, it seems probable that the repulsive force would overwhelm the effects of ordinary, attractive gravity and cause the universe to embark on a phase of violent runaway inflation. In the minutest fraction of a second a submicroscopic region of space would swell up exponentially to cosmic proportions, doubling its size every 10^{-35}s or so. This headlong distension would continue until at some point the universe would flip into its other 'frozen' phase, in which the forces separate and the repulsion disappears. In the absence of its driving force the exponential growth would come to a shuddering halt, amid a burst of heat, and the universe would return to the more conventional activity of gradually decelerating expansion, the remnants of which persist today.

The inflationary universe scenario solves several major cosmological problems at a stroke. For example, it explains why the universe is so uniform. Any initial irregularities would be drastically 'diluted' by the huge inflation. A bubble of space no larger than a proton might inflate to many times the present observable volume of the universe. Irregularities in the universe on proton scales and above would thus be stretched to insignificance within our observable universe.

Inflation would also explain the otherwise miraculous balance between the explosive vigour of the big bang and the gravitating power of the cosmic material. According to Guth, any excess or shortfall in the expansion rate is wiped out when the exponential inflation takes over, which also has the effect of inhibiting the formation of monster black holes in the primeval phase. By the time the universe exits from exponentiation, the deviation from matching will have been reduced to very nearly zero. (Although not strictly zero, evidently. Galaxies can still form.)

Finally, inflation also solves the horizon problem. Regions of the universe on opposite sides of the sky usually regarded as causally disconnected were, in fact, momentarily in causal contact before the inflationary phase. Everything we observe (and much more) was squeezed into a microscopic region of space at the start of the inflation. The horizon does not exist (at least not where we thought). Its prediction was based on the assumption that the universe had decelerated smoothly ever since the creation, and ignores the period of exponential growth.

In spite of its success in providing a neat explanation for several of cosmology's longstanding puzzles, the inflationary scenario is not without difficulties. The chief snag is known as the 'graceful exit' problem. In order for inflation to work its magic, the period of exponential growth must persist for long enough for the universe to swell by very many powers of ten. The sudden colossal expansion causes the temperature to drop more or less instantaneously to very nearly absolute zero. There seems to be nothing to stop the 'freeze-out' from happening immediately, thereby arresting the inflation before it really gets under way.

In an early version of the theory, Guth suggested that perhaps the universe underwent a period of so-called supercooling. This is a phenomenon known to physicists in more mundane contexts. Water, for example, if pure, can be carefully cooled below freezing point without solidifying. A slight disturbance, however, will cause a sudden transition to ice. In the cosmic case, supercooling could enable the universe to hang in the high-temperature (unified force) phase for a sufficient time for the inflationary period to proceed. The trouble comes when the freeze-out occurs. It seems likely that 'bubbles' of the new ('frozen') phase would appear at random and start to grow at the speed of light. Inside the bubbles inflation is absent, the energy of the inflationary growth being transferred instead to the bubble walls. Eventually the bubbles become large enough to intersect. The collisions between the highly energetic bubble walls would then introduce a great deal of turbulence and irregularity, the very features that the scenario was designed to ameliorate.

Work continues on how to avoid this tangled mess, wrecking the benefits of the inflation. One idea is that the bubbles may grow large enough to encompass the entire universe and much else, so that our observed cosmos would be a relatively uniform and quiescent oasis in a universe that, on a very large scale, is irregular and turbulent. Another suggestion is that, rather than supercooling followed by bubbles, the freeze-out could simply be a very sluggish process, allowing a relatively long period of inflation to occur before the phase transition could catch up. Many of these details are highly model-dependent, and it is too soon to say whether the graceful exit problem will be satisfactorily solved.

In spite of these technical difficulties the broad success of the inflationary scenario has endeared it to many physicists and cosmologists. If correct, it means that the universe need not have been created in a very special, ordered state after all. Initial gravitational irregularities were wiped out by inflation, while the subsequent expansion allowed the initially structureless, cosmic material to evolve complex structure and organization. Thus the origin of all complex cosmic order may be explained as the result of perfectly natural processes.

5. *God.*

If the grand unified theories fail, and if the anthropic argument is rejected, then the highly uniform nature of the universe on the large scale might be advanced as evidence for a creative designer. It would, however, be negative evidence only. No one could be sure that future progress in our understanding of the physics of the early universe might not uncover a perfectly satisfactory explanation for an orderly cosmos. Just as the formation of complex, ordered structures like the solar system was once attributed to a Deity, but then subsumed within the province of standard astrophysics, so too may the mysteries surrounding the large scale cosmic order come to be understood in purely natural, rather than supernatural, terms.

Our conclusion must be that there is no positive scientific evidence for a designer and creator of cosmic order (in the negative entropy sense). Indeed, there is strong expectation that current physical theories will provide a perfectly satisfactory explanation of these features.

There is, however, more to nature than its mathematical laws and its complex order. A third ingredient requires explanation too: the so-called 'fundamental constants' of nature. It is in that province that we find the most surprising evidence for a grand design.

By fundamental constants, physicists mean certain quantities that play a basic role in physics and which have the same numerical value everywhere in the universe and at all moments in time. A few examples will suffice to illustrate the idea. An atom of hydrogen is the same in a distant star as it is on Earth. It has the same size, mass and internal electric charges. But the values of these quantities are totally mysterious to us. Why is the proton in the hydrogen atom 1836 times as heavy as the electron? Why that number? Why are their electric charges what they are rather than some other value?

All nature's forces contain numbers like this that determine their strength and range. It may be that one day we shall have a theory that explains these numbers in terms of a more fundamental idea. Be that as it may, the actual values which the quantities assume turn out to be of crucial significance for the structure of the physical world.

Let us look at a simple example due to Freeman Dyson. The nuclei of atoms are held together by the strong nuclear force whose origin lies with the quarks and gluons described in Chapter 11. If the force were weaker than it is, atomic nuclei would become unstable and disintegrate. The simplest compound nucleus is that of deuterium (heavy hydrogen), consisting of a proton stuck to a neutron. This pair is glued by the strong nuclear force, but only tenuously. The link would be broken by quantum disruption if the nuclear force were only a few per cent weaker. The effect would be dramatic. The sun, and most other stars, use deuterium as a link in a chain of nuclear reactions to keep shining. Remove dueterium and either the stars go out, or they find a new nuclear pathway to generate their heat. Either way, they would be obliged to alter their structure drastically.

Equally dire consequences would ensue if the nuclear force were very slightly stronger. It would then be possible for two protons to overcome their mutual electric repulsion and stick together. During the big bang protons were much more abundant than neutrons. When the primeval material cooled, the neutrons sought out protons to stick to. The resulting deuterium soon underwent further synthesis to form the element helium. But the residue of unmatched protons remained unscathed to form the raw material of stars. If these protons could stick together in pairs, one member of each pair would decay to a neutron, converting the di-proton into deuterium and thence helium. So in a world where the nuclear force was a few per cent stronger, there would be virtually no hydrogen left over from the big bang. No stable stars like the sun could exist, nor could liquid water. Although we do not know why the nuclear force has the strength it does, if it did not the universe would be totally different in form. It is doubtful if life could exist.

What impresses many scientists is not so much the fact that alterations in the values of the fundamental constants would change the structure of the physical world, but that the observed structure is remarkably sensitive to such alterations. Only a minute shift in the strengths of the forces brings about a drastic change in the structure.

Consider as another example the relative strengths of the electromagnetic and gravitational forces in matter. Both forces play an essential role in shaping the structure of stars. Stars are held together by gravity, and the strength of the gravitational force helps determine such things as the pressure inside the star. On the other hand, energy flows out of the star by electromagnetic radiation. The interplay of these two forces is complicated, but reasonably well understood. Heavy stars tend to be brighter and hotter, and have no difficulty in transporting the energy generated in the core to the surface in the form of light and heat radiation. Light stars, however, are cooler, and their interiors cannot divest themselves of energy fast enough by means of radiation alone: they must be assisted by convection, which causes the surface layers to boil.

These two types of stars — hot and radiative or cool and convective — are known respectively as blue giants and red dwarfs. They delimit a very narrow range of stellar masses. It so happens that the balance of forces inside stars is such that nearly all stars lie in this very narrow range between the blue giants and red dwarfs. However, as pointed out by Brandon Carter,[2] this happy circumstance is entirely the result of a remarkable numerical coincidence between the fundamental constants of nature. An alteration in, say, the strengths of the gravitational force by a mere one part in 10^{40} would be sufficient to throw out this numerical coincidence. In such a world, all stars would then either be blue giants or red dwarfs. Stars like the sun would not exist, nor, one might argue, would any form of life that depends on solar-type stars for its sustenance.

The list of numerical 'accidents' that appear to be necessary for the observed world structure is too long to review here. (The reader is referred to my book *The Accidental Universe* for a complete discussion.) Opinions differ among physicists as to the significance of these coincidences. As with the apparently contrived initial conditions of the universe, recourse could be made to anthropic considerations and hypotheses of multiple-universes in which, for some reason, the fundamental constants assume different values. Only in those universes where the numbers come out just right would life and observers form.

Alternatively the numerical coincidences could be regarded as evidence of design. The delicate fine-tuning in the values of the constants, necessary so that the various different branches of physics can dovetail so felicitously, might be attributed to God. It is hard to resist the impression that the present structure of the universe, apparently so sensitive to minor alterations in the numbers, has been rather carefully thought out. Such a conclusion can, of course, only be subjective. In the end it boils down to a question of belief. Is it easier to believe in a cosmic designer than the multiplicity of universes necessary for the weak anthropic principle to work? It is hard to see how either hypothesis could ever be tested in the strict scientific sense. As remarked in the previous chapter, if we cannot visit the other universes or experience them directly, their possible existence must remain just as much a matter of faith as belief in God. Perhaps future developments in science will lead to more direct evidence for other universes, but until then, the seemingly miraculous concurrence of numerical values that nature has assigned to her fundamental constants must remain the most compelling evidence for an element of cosmic design.

14. Miracles

> 'God never wrought miracle to convince atheism, because his ordinary works convince it.'
>
> Francis Bacon

> 'There is not to be found, in all history, any miracle attested by a sufficient number of men, of such unquestioned good-sense, education, and learning, as to secure us against all delusion in themselves.'
>
> David Hume

However persuasive they may seem, arguments for the existence of God based on cosmology or suggestions of design in the natural world, are at best indirect. Some people, though, claim that God's activity can also be witnessed directly in the physical world, through miracles. All the world's major religions possess a folk lore about miracles. The Bible contains many such accounts, and even today reports of miracles are not uncommon.

In trying to assess the significance of such evidence the first problem is to decide exactly what is meant by a miracle, and there is by no means unanimous agreement on this. 'A miracle of modern science' conveys the impression of something unusual and spectacular, but nobody would suggest the word is being used literally in such a case. Aquinas defined a miracle as something 'done by divine power apart from the order generally followed in things'. In modern jargon this means a violation of the laws of nature produced by God. In other words, God intervenes directly in the operation of his world and changes something by 'breaking the rules'. If such events could be definitively verified they would indeed provide powerful evidence both for God's existence and his concern for the world.

Sometimes however a miracle is taken to imply something weaker. Many a 'miraculous escape' has convinced a lucky person of God's benevolence. The lone survivor of a plane crash may regard his deliverance as a miracle, even though the same event led to the pointless destruction of his fellow passengers.

This 'guardian angel' interpretation of extraordinary events belongs to quite a different category from the explicit violation of natural laws. Nobody suggests that surviving a plane crash must entail a suspension of the laws of physics. Such events are merely remarkable coincidences within the normal operation of physical processes. The proverbial parachutist with a malfunctioning parachute who lands in a haystack is simply lucky to have fallen where he did. No direct divine intervention seems to be involved.

Those who choose to read divine significance into improbable coincidences and lucky escapes are simply giving a theistic interpretation of straightforward, if unusual, natural events. But however convinced the lucky person himself may be that 'the gods are smiling on him', it is hard to make an objective case for the existence of God from events of this sort. The man who wins a fortune on the football pools may reflect on the fact that, purely by the rules of the game, someone will win. And the soldiers who, claiming God's help, slaughter their adversaries in battle, might ask themselves where God was when the enemy soldiers needed him.

Believer: In my opinion, miracles are the best proof that God exists.

Sceptic: I'm not sure I know what a miracle is supposed to be.

Believer: Well, something extraordinary and unpredictable.

Sceptic: The fall of a large meteorite, or the eruption of a volcano is extraordinary and unpredictable. You aren't suggesting they are miraculous surely?

Believer: Of course not. Such phenomena are natural events. Miracles are *supernatural*.

Sceptic: What do you mean by supernatural? Isn't it just another word for miraculous? (Consults Oxford dictionary.) It says here: 'Supernatural. Outside the ordinary operation of cause and effect'. Hmm. It all depends on what you mean by 'ordinary'.

Believer: I would say ordinary meant familiar or well understood.

Sceptic: A dynamo or a radio would have been regarded as miraculous by our ancestors, who were not familiar with electromagnetism.

Believer: I agree they probably would have regarded these devices as miraculous, but erroneously, for we know they operate according to natural laws. A truly supernatural event is one whose cause cannot be found in any natural law, *known* or *unknown*.

Sceptic: Surely that is a useless definition? How do you know which laws might be unknown? There may be totally bizarre and unexpected laws that we may simply happen not to have stumbled across. Suppose you saw a rock float in the air. Would you regard that as a miracle?

Believer: It depends . . . I would have to be sure there was no illusion, or trickery.

Sceptic: But there may be natural processes that produce super illusions that nobody would suspect.

Believer: Or perhaps all our experience is an illusion and we might as well give up discussing anything?

Sceptic: O.K., let's not take that route. But you still can't be sure that some quirky magnetic or gravitational effect isn't making the rock levitate.

Believer: But it's easier to believe in God than outlandish magnetic phenomena. It's all a question of credibility.

Sceptic: Ah! So by a miracle you really mean 'something caused by God'?

Believer: Absolutely! Though he may sometimes use human intermediaries.

Sceptic: Then you cannot present miracles as evidence for God, or your argument is circular. 'Miracles prove the existence of an agency which produces miracles.' What it really boils down to, as you admitted, is belief. You have to believe in God already for miracles to have any meaning. Apparently miraculous events in themselves cannot prove the existence of God. They might be freak natural events.

Believer: I concede that levitating rocks are dubious from the miracle point of view, but consider some of the famous miracles: Jesus' feeding of the multitude, for example. You can't tell me any sort of natural law would duplicate loaves and fishes!

Sceptic: But what possible reason can you have for believing a story written hundreds of years ago by a lot of superstitious zealots with a vested interest in promoting their own brand of religion?

Believer: You are remarkably cynical. Taken in isolation, the loaves and fishes story is nothing. You have to see it in the context of the whole Bible. It was not the only miracle reported there.

Sceptic: Remind me of another.

Believer: Jesus walked on the water.

Sceptic: Levitation! I thought you'd dismissed that sort of miracle as 'dubious'.

Believer: For a rock yes, for Jesus, no.

Sceptic: Why not?

Believer: Because Jesus was the Son of God and so possessed supernatural powers.

Sceptic: But you're begging the question again. I don't believe Jesus had supernatural powers. If he did walk on water I would rather suppose it to have been a freak natural event. However, I don't believe the story anyway. Why should I?

Believer: The Bible has been a source of inspiration to millions. Don't dismiss it lightly.

Sceptic: So have the works of Karl Marx. I wouldn't believe any account of his about miracles either.

Believer: You may refuse to accept the word of the Bible, but you can't dismiss the claims of hundreds of people who have experienced miracles even in recent years.

Sceptic: People claim all sorts of things: meetings with aliens, teleportations, clairvoyance. Only a fool or a madman would listen to such nonsense.

Believer: I concede that many wild and fanciful claims are made, but the evidence for faith healing is compelling. Think of Lourdes.

Sceptic: Psychosomatic! Let me quote you: 'It's all a question of credibility.' I agree. Surely it's easier to believe in a few freak medical events than to invoke a Deity?

Believer: You can't debunk all miracles as psychosomatic. What does that term mean anyway? It's just a euphemism for 'medically inexplicable'. Why should so many people be so convinced by miracles if they were just natural freaks?

Sceptic: It's all a hang-over from the age of magic. Before the rise of science, or the great world religions, primitive peoples believed that almost anything which happened was caused by magic — the action of some minor god or demon. As science explained more and more, and religion groped towards the idea of one God, so the magical explanations became moribund. But a vestige lives on.

Believer: You're not suggesting that Lourdes pilgrims are demon worshippers!

Sceptic: Not overtly. But their belief in faith healing differs very little, maybe not at all, from the beliefs concerning African witch-doctors, or spirit contacts, for example. Atavistic superstitions from the age of magic have simply been institutionalized by the great religions. Talk of miracles is just sanitized magic-mongering.

Believer: There are powers of good and evil. They manifest themselves in many ways.

Sceptic: And do you take evil supernatural events as evidence for God too? Does he also wield evil powers?

Believer: The relation between good and evil is a delicate theological subject. There are many shades of opinion about your questions. Man's wickedness can act as a channel for evil, whatever its ultimate origin.

Sceptic: So you would not necessarily make God responsible for the so-called occult powers, if they exist?

Believer: Not necessarily, no.

Sceptic: So there are at least two types of supernatural events, then: those that originate with God — what you have called miracles — and the nasty ones — the black arts, shall we say — the origin of which is controversial. Then there would be the neutral ones, I suppose. Like psychokinesis and precognition? It all sounds a bit complicated to me. I'd rather believe that all these topics are just primitive fantasies, a relic of the age of magic, a vestige of polytheism. Your belief in miracles is just the respectable end of a spectrum of neurotic primeval superstitions, and quite unworthy of a God of the majesty and power that you describe.

Believer: It seems to me to be not at all unreasonable to suppose that supernatural powers exist, and can be manipulated in a variety of ways, for good or evil. Faith healing is the good side.

Sceptic: And provides evidence for God?

Believer: I believe so.

Sceptic: What about the failures, the unfortunate ones who don't respond to the healing? Doesn't God care about them? Or does his power waver occasionally?

Believer: God moves in mysterious ways, but his power is absolute.

Sceptic: That's just a platitudinous way of saying you don't know. And if God's power is absolute why does he need miracles anyway?

Believer: I don't understand.

Sceptic: An omnipotent God, who rules the entire universe, and who can make anything happen, has no need of miracles. If he wants to avoid somebody dying of cancer he could prevent them contracting the disease in the first place. In fact, I would regard a miracle as evidence that any God had lost control of the world, and was clumsily trying to patch up the damage. What is the point of God doing all these miracles?

Believer: Through miracles, God demonstrates his divine power.

Sceptic: But why is he so obscure about it? Why does he not write a clear proclamation in the sky, or turn the moon tartan, or something else utterly incontrovertible? Better still, why not avert some major natural disaster, or prevent the spread of devastating epidemics? However wonderful a few cures at Lourdes may be, the stock of human misery is enormous. I repeat, the miracles you describe seem unworthy of an omnipotent God. Levitation, multiplying fishes — they have the air of a cosmic conjuring act. Surely they are just products of puerile human imagination?

Believer: Perhaps God *is* averting disasters all the time.

Sceptic: That's no reply! Anyone could claim the same. Suppose I say that by pronouncing an incantation each morning I prevent world war, and cite as evidence the fact that world war has indeed not broken out? In fact a group of UFO buffs claim just that.

Believer: Christians believe that God continually holds the world in being, so in a sense everything that happens is a miracle, and all this talk of distinction between natural and supernatural is actually a red herring.

Sceptic: Now you're shifting ground. You seem to be saying God *is* nature.

Believer: I'm saying that God causes everything in the natural world, though not necessarily in the temporal sense. He doesn't just set the whole thing going and then sit back. God is outside the world, and *above* the laws of nature, sustaining all of existence.

Sceptic: It seems to me we have a semantic quibble here. Nature has a beautiful set of laws and the universe runs along a pathway of evolution mapped out by those laws. You describe exactly the same thing in theistic terms by talk of 'upholding'. Your God is only a mode of speaking, surely? What does it mean to say God upholds the universe? How is that different from simply saying that the universe continues to exist?

Believer: You cannot be content with the bald fact that the universe exists. It must have an *explanation*. I believe God is that explanation, and his power is employed at every moment sustaining the miracle of existence. In most cases he does this in an orderly way — what you would call the laws of physics — but from time to time he departs from this order and produces dramatic events as warnings or signs to human beings, or to assist the faithful, such as when he parted the Red Sea for the Hebrews.

Sceptic: What I find hard to understand is why you think that this supernatural miracle-worker is the *same* as the being who created the universe, who answers prayers, who invented the laws of physics, who will sit in judgement and so on. Why can't all these individuals be different supernatural agents? I should have thought that with so many miracles apparently supporting many different and conflicting religions, a believer in miracles would be obliged to concede the existence of a whole host of supernatural beings in competition.

Believer: One God is simpler than many.

Sceptic: I still don't see how so-called miraculous events, however remarkable, can be regarded as evidence of God's existence. It seems to me you are simply exploiting the fairy godmother instinct we all have, turning 'Lady Luck' into a real being and calling her God. How can you take these 'miracles' seriously?

Believer: I don't find anything incredible in God, who is creator of all, manipulating material objects. Compared to the miracle of his universe, what is so remarkable about God parting the Red Sea?

Sceptic: But you are still basing your argument on the assumption that God exists. I agree that if a God of the sort you describe — infinite, omnipotent, benevolent, omniscient, and so on — does exist, the Red Sea would be a triviality for him. But how do we know he does exist?

Believer: It's all a question of faith.

Sceptic: Precisely!

This inconclusive dialogue I hope brings out the essence of the conflict between science and religion when it comes to supernatural matters. The religious person, who is comfortable with the notion of God's activity and sees God's work all around him every day, finds nothing incongruous about miraculous events because they are simply another facet of God's action in the world. In contrast the scientist, who prefers to think of the world as operating according to natural laws, would regard a miracle as 'misbehaviour', a pathological event which mars the elegance and beauty of nature. Miracles are something that most scientists would rather do without.

The evidence for miracles is, of course, highly controversial. If they were to be accepted solely on the basis of existing testimony, there would be no good reason not to accept a host of other claims (UFOs, ghosts, spoon-bending, mind-reading, for instance) that seem equally well attested. But even if a scientist is persuaded to accept miracles, there can be no real dividing line between the miraculous and what is now known as the paranormal.

There is a huge and rising interest in paranormal phenomena, from metal-bending to ESP. Very few paranormalists imbue their subject with theological connotations. The phenomena, even in cases of healing, are regarded as 'Godless miracles'. The primitive beliefs and hysteria which accompany many excursions into the paranormal serve to degrade religion. A well-known Sunday newspaper supplement once compared Jesus Christ with Uri Geller. Unfortunately, many reported miracles do have the flavour of music-hall stunts. St. Joseph of Cupertino, for example, is said to have so embarrassed his holy brothers by his tendency to float in the air at worship that he was confined to his cell for the purpose of mass!

It is interesting to note that many of the symbols of alleged supernatural religious events have reappeared among the modern UFO cults. Take, for example, the stories of witnesses who claim to have been abruptly cured of some long-standing medical complaint after an encounter with UFO occupants, or occasionally by merely sighting a UFO itself.

Levitation also plays a prominent part. Speeding serenely and silently through the sky, the ephemeral flying saucers are powered, we are assured, not with the aid of crude rockets or brute force motors, but by neutralization of the Earth's gravity. Sometimes the ufonauts themselves float about weightless at ground level.

Clearly, aerial phenomena, levitation and healing powers are deep-rooted in the human psyche. In the age of magic they were prominent and overt. With the development of organized religion, they became refined and submerged, but the strong primeval element has never been far below the surface. Now, with the decline of organized religion, they have re-surfaced again in technological guise, employing the language of spacecraft and pseudoscience, of mysterious force fields and mind over matter — a polyglot synthesis of primitive superstition and space-age physics.

Miracles have always constituted the showbizz end of religion, and stood uneasily beside the other alleged paranormal phenomena, much of which, like diabolism, seems most unsavoury. The believer has the doubly-difficult job of first persuading the sceptic that the phenomena really occur, which is a daunting task given the dubious nature of most testimonies, and then convincing him that miracles are in any direct way connected with God. This means either accepting all supernatural events (even the unpleasant ones) as the work of God, or somehow establishing a clear distinction between God's miracles and the rest. And in an age when ESP is as familiar as ABC, most of those who are convinced of miracles would rather put their money on mind-power than God-power.

15. The end of the universe

Sic transit gloria mundi

If the universe has been designed by God, then it must have a purpose. If that purpose is never achieved, God will have failed. If it is achieved, the continuation of the universe will be unnecessary. The universe, at least as we know it, will come to an end.

Religions differ greatly in their conception of the moment and manner of the cosmic demise. Some warn of imminent catastrophe, of a world overtaken by apocalytic destruction when the sinful will be judged sternly. Others teach about a forthcoming Kingdom of Heaven which will replace the harsh and uncertain world we now know. Some Oriental religions incline towards a cyclical system, where the end of this world presages the rebirth of another, similar world.

What has modern science to say about the end of the universe?

In Chapter 2 it was explained how the second law of thermodynamics operates inexorably to reduce the organization of the universe to chaos. Everywhere we look, in every corner of the cosmos, entropy is rising irreversibly and the vast stock of cosmic order is slowly but surely being depleted. The universe seems destined to continue crumbling, running down towards a state of thermodynamic equilibrium and maximum disorder, after which nothing further of interest will happen. Physicists call this depressing prospect 'the heat death'; it has been discussed for over a century.

The second law of thermodynamics is so fundamental to all of physics that few physicists would question its validity. As we saw in Chapter 9, the second law is responsible for imposing a time asymmetry on the world that gives it a distinction between past and future. Violating the second law would be tantamount to reversing the direction of time.

The second law tells us nothing, however, about the nature of the cosmic catastrophes that will drive the universe towards its end state of maximum disorder. In the last thirty years, with the rapid development of modern astronomy, it has become possible to fill in some of the details of the events that will irresistibly destroy the complex organization and elaborate activity of the world about us.

As far as our local region of the universe is concerned the fate of the Earth is intimately linked with the fate of the sun. Earth-life feeds on sunlight, and any major disruption to the sun's present quiescence will spell disaster. There is no lack of possible solar twitches that could render the Earth uninhabitable. Any alteration in the sun's constant heat output could upset the Earth's delicate climatic balance and plunge us into a catastrophic ice age. Changes in the magnetic patterns of the solar system connected with the so-called solar wind — a steady stream of particles from the sun's surface — might bring about equally drastic changes. The explosion of a nearby star could drench us in lethal radiation, or the passage of a black hole through the solar system might rock the planets in their orbits.

But assuming the Earth escapes all these unpleasant possibilities, it is clear that things cannot continue as now 'for ever and ever'. The prolific radiation of energy by the sun has to be paid for in nuclear fuel, and eventually the fuel reserves will start to run out. Astrophysicists estimate that this will not happen for another four to five billion years, which may seem a vast length of time. However, by comparison the age of the universe is eighteen billion years, while the sun is already 4½ billion years old, making it comfortably middle aged.

As the fuel runs low, so the sun will swell up, turning into the sort of star that astronomers call a red giant. The core of the sun, struggling desperately to keep up energy production, will shrink further and further until quantum effects intervene to stabilize it. At this stage the sun may have become so distended that the inner planets will have been engulfed, the Earth's atmosphere stripped away and the solid rocks melted or even vaporized. Thereafter the sun will embark on a new and erratic career, in which the nuclear burning of hydrogen fuel, so prolific today, will be replaced by the less efficient burning of helium, and thence by heavier and heavier elements, in the manner described in Chapter 13.

When finally all fuel is exhausted, the sun will consist of moderately heavy elements such as iron. Any further fusion of nuclei would not result in the release of energy. Iron is the most stable nuclear form, and according to the second law of thermodynamics, all systems seek out their most stable state. During this phase the sun's central temperature will have risen steadily towards a billion degrees. Now, with all fuel spent, the internal pressure will falter and gravity will take command. The stricken sun will start to contract under its own weight, crushing the material within it so violently that the density will rise to a million grams per cubic centimetre. The shrunken, burnt-out sun will be reduced to the size of the Earth and sit, inert, for countless billions of years slowly fading and cooling to end its career as a black dwarf star.

The same pattern of instability, distension, fuel starvation and collapse will be repeated throughout our galaxy and in every other galaxy. One by one the stars will burn their way through the nuclear cycle until they can no longer hold up their own weight against the remorseless force of gravity.

Some stars will die in more spectacular fashion, as supernovae, blowing themselves to smithereens as their cores implode catastrophically and release tremendous energy. The remnants of the lighter of these kamikaze stars will consist of diffuse debris surrounding a chunk of ultra-crushed matter in which the equivalent of one solar mass is compressed in a spherical volume only a few miles across. So immense is the gravity of such an object that a teaspoonful of matter there would weigh more than all the continents weigh on Earth. It is a grip too fierce for even the atoms to withstand, and they are forced to crumple inwards to form a sea of pure neutrons. These neutron stars are familiar to astronomers who find them amid the debris of past supernova explosions.

Heavier dead stars are unable to stabilize themselves in the face of gravity's crushing power even by converting to a ball of neutrons. Such stars continue to shrink at an escalating rate, to end their days as black holes.

The cosmologist Edward Harrison describes the slow degeneration of the universe in the following graphic terms:

> The stars begin to fade like guttering candles and are snuffed out one by one. Out in the depths of space the great celestial cities, the galaxies, cluttered with the memorabilia of ages, are gradually dying. Tens of billions of years pass in the growing darkness. Occasional flickers of light pierce the fall of cosmic night, and spurts of activity delay the sentence of a universe condemned to become a galactic graveyard.[1]

In their search for the highest entropy state, physical systems explore some bizarre pathways. The organization of our galaxy will begin to falter as the stars inexorably burn out. For stars like the sun this takes several billion years, and during this time new stars will continue to form from the interstellar gases. Smaller stars may take thousands of times longer to die. Eventually, however, the ordered energy locked up in the stars will have been spread chaotically around the universe in the form of radiation, and the galaxy will dim and cool. A similar fate will befall other galaxies.

As for the dead stars themselves, there is plenty of further activity in store, but the time-scale is greatly increased. As the burnt-out remnants mill around the galaxy, from time to time collisions will occur. Black holes will tend to swallow up any stars, or other material that they encounter, and if, as some astronomers believe, there is a huge black hole at the centre of our galaxy, it will grow progressively larger. The orbits of the stars will slowly decay due to the emission of gravitational radiation — wavelike ripples of space that sap the orbital energy of all massive objects. Over an immensity of time, the stellar remnants will tend to drift closer and closer to the galactic centre, eventually to sacrifice themselves to the insatiable monster hole. Some dead stars will escape this fate as a result of fortuitous encounters with other stars which knock them out of the galaxy altogether, to roam in solitary confinement in the vastness of intergalactic space.

For the stars that do escape, and for any gas or dust that avoids a black hole death, the reprieve is only temporary. If the grand unified theory is correct, the nuclear material of this cosmic flotsam is unstable, and will evaporate away after about 10^{32} years. The neutrons and protons decay into positrons and electrons, which then start to annihilate with each other and any further electrons. All solid matter thus disintegrates. The outcome of this carnage depends on the speed with which the universe is actually expanding. If the more rapid estimates are correct, the electrons and positrons will be swept apart by the expansion faster than they can run into each other, so complete annihilation will not occur: there will always be some particles left. Those that do annihilate produce gamma radiation, which itself slowly weakens with the cosmic expansion. In addition there will be neutrinos and heat radiation left over from the big bang. All these components will gradually cool towards the absolute zero of temperature, but at different rates. The matter (electrons and positrons) will cool faster than the radiation. Thus, although both approach absolute zero, so that their temperature difference decreases, there will nevertheless always be a finite temperature gap which, in principle, can act as a source of free energy (negative entropy). Thus, although the entropy of this highly run-down universe is close to its maximum value, it never quite attains it, so to this limited extent, a true heat death will never occur.

If the universe expands rather more slowly, the annihilation of electrons and positrons will be more favoured. Mutual destruction does not happen, however, as a result of simple chance collisions. Electric forces attract electrons to positrons, enabling them to form 'atoms' known as positronium. Calculations suggest that in a slowly expanding universe most of the particles will have formed positronium after 10^{71} years. But these are bizarre 'atoms' indeed, being thousands of billions of light years in size! The particles orbit around each other so slowly it takes a million years for them to move a single centimetre. The positronium is unstable, and very gradually these immense orbits decay by the emission of very low energy photons. After 10^{116} years most of the positronium will have collapsed and the particles come into contact, whereupon instant annihilation occurs. During the decay no less than 10^{22} photons will have been emitted by each positronium 'atom' — a huge increase in entropy.

As far as the black holes are concerned, they do not remain inert either. The quantum effects discussed briefly in Chapter 13 imply that the holes are not strictly black, but glow dully with heat radiation. For a solar-mass black hole the temperature is a pitiful ten billionths of a degree above absolute zero, while for superholes it is even less. So long as the background temperature of the universe remains above this, the holes will continue to grow very slowly by heat absorption. Some activity will still occur as holes collide with other objects or other holes, and any rotating holes will gradually slow down as their spin is dissipated. But by far the most drastic change will start to occur when the temperature of space eventually drops below the temperature of the holes.

A black hole that is hotter than its environment will tend to lose heat and hence energy, which will cause it to shrink. This in turn slightly raises the temperature causing the emission to accelerate. The hole thus embarks upon the slippery slope of runaway evaporation. Over the aeons the rate of shrinkage escalates until, after perhaps 10^{108} years, even huge black holes, initially weighing many galactic masses, will have been reduced to nothing.

Nobody knows how a black hole finally dies, but it seems likely that it will shrivel to microscopic dimensions, becoming so hot that it begins to create matter. At that stage it has only a few billion years to live. Eventually the hole probably explodes amid a shower of gamma rays leaving no remains of its erstwhile existence.

These studies suggest a dismal fate for the universe we now know, so full of splendour and activity. Though the immensity of the times involved are beyond human imagination (recall that 10^{100} is one followed by one hundred zeros) there seems little doubt that all the currently-observed structures are destined to pass away eventually, leaving only cold, dark, expanding, near-empty space, populated at an ever-decreasing density by a few isolated neutrinos and photons, and very little else. It is a scenario that many scientists find profoundly depressing.

There is, however, an alternative fate. In arriving at the above conclusions the assumption was made that the universe will continue to expand for ever more. This is not clear. It is known that the rate of expansion is steadily falling, as gravity restrains the dispersal of the galaxies, and some astronomers believe that it will one day come to a halt. Whether or not this is the case depends on the gravitating power of the universe, which in turn depends on the density of material. As this includes unseen material such as neutrinos and black holes, as well as unseen energy such as gravity waves, it is almost impossible to assess what the overall density might be.

If the expansion does come to a halt, the universe will not remain static, but will start to contract in a motion that is the time reflection of its expansion phase. At first the contraction will be slow, but over many billions of years the pace will accelerate. Galaxies that are now receding from one another will start to approach instead, gathering speed all the time. The stage is set for a monstrous cataclysm.

When the universe has shrunk to one hundredth its present size, the compression effect will have elevated the temperature to the boiling point of water, and the Earth (if it has survived the sun's death throes) will become uninhabitable. An observer would no longer be able to discern individual galaxies, for these would by now have begun to merge with each other as the intergalactic spaces close up. Further shrinkage will raise the temperature to the point where the sky itself will begin to glow like a furnace, and the stars, embedded in this white hot space, will start to boil, then explode.

The pace of events now quickens. All structures are vaporized and their atoms dispersed. In a few hundred thousand years — no more — the very nuclei are themselves smashed to pieces in the escalating temperatures. The time scale of events now becomes frenetic. The universe is shrinking appreciably in minutes, then seconds, then microseconds, as the accumulating gravity turns the cosmic contraction into an uncontrolled implosion. This is the 'big crunch'.

The awesome nature of these events inspired the poet Norman Nicholson to write these words:

> And if the universe
> Reversed and showed
> The colour of its money;
> If now observable light
> Flowed inward, and the skies snowed
> A blizzard of galaxies,
> The lens of night would burn
> Brighter than the focussed sun,
> And man turn blinded
> With white-hot darkness in his eyes.[2]

The universe is now microseconds away from death. The big crunch is like the big bang in reverse. Nuclear particles break apart into quarks, all manner of subnuclear splinters are created for a fleeting instant but, in the twinkling of an eye, the entire universe shrivels into less than the space of an atom, whereupon spacetime itself disintegrates.

Many physicists believe that the big crunch represents the end of the physical universe. Just as they believe the universe — all space, time and matter — came into existence in a big bang, so they believe it goes out of existence in a big crunch. This is total annihilation. Nothing is left. No places, no moments, no things. There is a final 'singularity' as all of existence succumbs to the infinite destructive power of gravity, and then no more. Gravity, the midwife of the cosmos, is also its undertaker.

Not all scientists are prepared to accept this spectacular demise of the universe however. Some argue that unknown physical forces will cause the big crunch to stop at some fantastic density, causing the universe to 'bounce' back out again into another cycle of expansion and recontraction, followed by another and another, and so on, *ad infinitum*. This is the oscillating universe already mentioned in Chapter 12. Only further work on ultra-high energy physics is likely to settle the issue.

Although science provides a variety of scenarios for the fate of the universe, they all involve the demise of the universe as we know it today. To that extent they coincide with most religious eschatology. However, the time scales concerned are so unimaginably immense that it is impossible to relate the death of the cosmos to human activities. If any conscious creatures exist in a future so remote that the present epoch would seem indistinguishable from the creation, they will not be human beings. Trillions of years of evolution and technological advancement will take care of that.

In the first place the development of 'artificial' intelligence may well imply that man will relinquish his intellectual supremacy in favour of thinking machines. Indeed, this is already happening in a limited sense. With the oceans of time available for future innovation, there seems to be no reason why machines cannot achieve and surpass anything of which the human brain is capable. And because there is no size limitation for such machines, it is not hard to envisage monster artificial superbrains with an intellectual power utterly beyond anything we can now comprehend. Moreover, the ability of electronic devices to transfer information directly to each other opens the way to a synthesis of machine brains. One can imagine an elaborate network of radio communication throughout the universe connecting together countless sedentary superbrains into a single superduperbrain.

Advance in genetic manipulation could give a new twist to the concept of the thinking machine. Until now, the development of biological intelligence has been at the mercy of natural evolutionary forces, but as we gain control over the molecular structures that determine our physical and mental characteristics, so it will become possible to modify existing organisms or even invent new ones. This is already done in a limited way by cross-breeding and mutation inducement. There seems to be no fundamental reason why the day will not come when brains can be 'grown' to order through molecular engineering. The distinction between natural and artificial intelligence will then disappear. These superior brains, created by man, could either be regarded as genetically manipulated biological organisms or as advanced computers using organic rather than solid-state hardware. It is even possible to imagine a symbiosis of the two, where organic brains can 'plug in' to solid-state circuitry, or in which future superchips are implanted in brains as a sort of 'booster kit'. Alternatively, it may make sense to use organic components in place of some semiconductor crystals in more conventional thinking machines. Of course, nobody is suggesting that any of these possibilities is even remotely feasible in the foreseeable future, but can we honestly believe that they will not be achieved in, say, a million years of scientific research? A billion? A trillion? Science, remember, is only a couple of centuries old.

A related issue concerning the remote future of the universe and its inhabitants is the interesting question of whether there exists any limit to the degree of control that intelligent creatures can exercise over the natural world. The universe we see has been shaped by the great cosmic forces, from the powerful nuclear interactions to the long-range effects of gravity. But we also see the rudiments of artificial environments: rivers diverted and dammed, forests created and destroyed, deserts tamed, mountains razed. Few areas of the Earth's surface show no evidence at all of man's activities. As technology and scientific understanding advance we can expect our descendants to gain control over larger and more complex physical systems. Freeman Dyson has envisaged advanced technological communities which drastically modify the structure of their planetary systems, creating a spherical shell of material around their star to trap and utilize the maximum available energy output. The level of technology required to dismantle planets may seem to lie forever in the realm of fantasy, but such a venture requires primarily time, money and resources — not skill.

We are therefore faced with an intriguing prospect. In a universe with virtually unlimited time available for technological enterprise, can we confidently rule out *anything* that is consistent with the laws of physics? During the last few thousand years humans have progressed from technology on the scale of hand tools a few centimetres in size to major engineering projects (bridges, tunnels, dams, cities) many miles in size. If that trend is extrapolated, even at a greatly diminished pace, the time will come when the whole Earth, then the solar system, and eventually the stars will be 'technologized'. The galaxy itself could be remodelled by manipulation, stars moved out of their orbits, created from gas clouds, or destroyed by engineered instabilities. Black holes could be formed or controlled at will as energy sources and/or disposal devices for the effluents of cosmic society.

And if galaxies, why not the universe?

Such extrapolations may be dismissed as absurd but they do raise an important philosophical point. What, if any, is the distinction between natural and artificial, between blind forces and intelligent control? This is a new angle on the controversy about free will and determinism.

When a system is brought under intelligent control, it still conforms to the laws of physics. There is no evidence (except arguably at the level of the mind-body interaction) that a major artificial construction violates any physical principles. It is true that a railway network or a nuclear power station, for example, would not arise spontaneously, but its construction still takes place within the framework of the laws of nature. And the resulting order thereby achieved is offset by the spiralling entropy generated in the construction process.

As discussed in Chapter 6, in the operation of the brain, there is a description in terms of physical laws at the hardware level, and an equivalent, consistent, description in terms of thoughts, sensations, ideas, decisions, and so on, at the software level. Likewise, to say that a system has become 'technologized' is not to deny the authority of physical law, but merely to use software language in describing its operation. There is no conflict, then, in a universe that evolves according to well-defined laws of physics but is nevertheless subject to intelligent control.

This is a thought-provoking conclusion. Those who invoke God as an explanation of cosmic organization usually have in mind a *supernatural* agency, acting on the world in defiance of natural laws. But it is perfectly possible for much, if not all of what we encounter in the universe to be the product of intelligent manipulation of a purely natural kind: within the laws of physics. For example, our galaxy *could* have been made by a powerful mind who rearranged the primeval gases using carefully placed gravitating bodies, controlled explosions and all the other paraphernalia of a space age astro-engineer. But would such a superintelligence be God?

The point is not a trivial one. God is usually conceived as the creator of the whole universe (including space and time) and not merely as a galactic architect. Clearly, no being who is obliged to operate *within* the physical universe, employing only pre-existing laws, can be considered as a universal creator. Yet suppose the powers of this super astro-engineer were extended to include all galaxies? Imagine that he can bend space and time using gravity.

Still, he would not be God if he could not actually create or destroy space and time. But the new physics gives a curious slant here. Given enough energy and resources, it is within the power of human beings to accumulate enough gravitating material to make a black hole. At the centre of the black hole, at the so-called singularity, space and time *are* destroyed. So even *we* can destroy spacetime.

Creating spacetime is harder. But can we be sure that it is truly *impossible* – completely ruled out by the laws of physics? We cannot. Indeed, in Chapter 3 some recent theories about the creation of 'bubbles' of space in the big bang were described. Moreover, what if space and time are after all eternal, contrary to the popular theory of a big bang creation? If space and time have always existed, it makes no sense to talk about the universe being created in time anyway. So

God's function within the universe would be restricted to fashioning and organizing matter, which he could perhaps accomplish entirely by natural means (we side-step some thermodynamic problems here). According to such a view God could be eternal, infinite and the most powerful being in the universe. He would not be omnipotent for he could not act outside the laws of nature. He would be the creator of everything we see, having made matter from pre-existing energy, organized it appropriately, set up the conditions necessary for life to develop, and so on, but he would not be capable of creation out of nothing (*ex nihilo*) as Christian doctrine requires. We might call this being a natural rather than a supernatural God.

What evidence do we have that such a natural God exists? Is the evidence better or worse than that for a supernatural God?

There are many mysteries about the natural world that would be readily explained by postulating a natural Deity. The origin of galaxies, for example, has no satisfactory explanation at present. The origin of life is another baffling puzzle. But we can conceive of both these systems being deliberately engineered by an intelligent super-being, without any violation of the laws of physics. However, such an explanation falls into the old trap of attributing to God anything that happens to be beyond the scope of today's scientific understanding. (The 'God of the gaps'.) Religious adherents have learned to their cost how perilous it is to point to a phenomenon and say 'That is evidence of God's work' only to find that scientific advances subsequently provide a perfectly adequate explanation. To invoke God as a blanket explanation of the unexplained is to invite eventual falsification, and to make God the friend of ignorance. If God is to be found, it must surely be through what we discover about the world, not what we fail to discover.

Nevertheless a natural, as opposed to a supernatural God fares better in the face of this argument. The hypothesis that a natural God created life, within the laws of physics, is at least known to be possible and consistent with our scientific understanding of the physical world, if only for the reason that the creation of life by man in the laboratory is a distinct (if remote) possibility.

How can we weigh the credibility of the two explanations for the origin of life (or any other highly ordered system): that life is the product of intelligent, but natural, manipulation by a superbeing, perhaps the supreme being (God), or that life is the end result of mindless self-organizing processes (like the appearance of ordered convection patterns in Jupiter's atmosphere)? Neither explanation is without difficulty.

The answer to this question must hinge on the extent to which we believe that mind is an important force in the universe. Most people are prepared to accept the sci-fi scenario of a remote future in which even greater regions of the universe come under intelligent control. In the countless trillions of years ahead of us, one can imagine that all the universe which we can now observe will be technologized. Why, then, is it so hard to imagine that such a superintelligence cannot have existed before us?

The conventional position is that intelligence only arises as the end product of a long sequence of evolutionary changes which successively increase the degree of organization of matter. In short, matter first, mind later. But is this inevitably so? Could mind be the more primitive entity?

There is a growing appreciation among scientists that neither mind, nor life, need be limited to organic matter. In a recent extremely speculative but thought-provoking book — *Life Beyond Earth* — the possibilities for extraterrestrial life are reviewed by the physicist Gerald Feinberg and the biochemist Robert Shapiro. They argue the case for life based on plasmas, electromagnetic field energy, magnetic domains in neutron stars and a variety of other bizarre systems. Now consciousness and intelligence are software concepts; it is only the pattern — the organization — that counts, not the medium for its expression. Taken to its logical conclusion, it is possible to imagine a supermind existing since the creation, encompassing all the fundamental fields of nature, and taking upon itself the task of converting an incoherent big bang into the complex and orderly cosmos we now observe; all accomplished entirely within the framework of the laws of physics. This would not be a God who created everything by supernatural means, but a directing, controlling, universal mind pervading the cosmos and operating the laws of nature to achieve some specific purpose. We could describe this state of affairs by saying that nature is a product of its own technology, and that the universe *is* a mind: a self-observing as well as self-organizing, system. Our own minds could then be viewed as localized 'islands' of consciousness in a sea of mind, an idea that is reminiscent of the Oriental conception of mysticism, where God is then regarded as the unifying consciousness of all things into which the human mind will be absorbed, losing its individual identity, when it achieves an appropriate level of spiritual advancement.

It is possible to go further than this. Recall that, according to some physicists at least, the mind has a special status in regard to the quantum factor. If mind can 'load the quantum dice' then a universal mind could, in principle, control everything that happens by directing the behaviour of every electron, every proton, every photon, and so on. Such an organizing power would escape our attention when we observe microscopic matter because the antics of any particular particle would still appear to be completely random. It is only in the collective behaviour of vast numbers of atoms that organization would be apparent, and we should proclaim the system to be mysteriously self-organizing. Such a picture of God might well be enough to satisfy most believers.

Many early religions involved a polytheistic scheme in which gods were graded according to their powers. This idea finds a counterpart in modern speculation about alien intelligences. Some writers have envisaged a hierarchy of intellectual and technological power, ranging from mankind upwards. One can envisage creatures whose capabilities are so great we could not distinguish their activities from nature itself. This hierarchy would involve a supreme being possessing the greatest power and intelligence. Such a being would fulfil many of the traditional requirements of God.

If we were persuaded that such a mind existed (and no amount of scientifically *possible* scenarios provide any actual evidence that it does), would this being be capable of preventing the end of the universe?

If the supreme being is constrained to act within the laws of physics (albeit the rather flexible laws permitted by the quantum theory) then the answer must be no. The second law of thermodynamics forbids anyone, however proficient their technology and however deep their understanding, to reverse the relentless rise in entropy.

It might be supposed that a being who could manipulate matter at the atomic level would be able to keep 'rewinding' the universe by restoring the flagging organization. This is actually an old idea explored by Maxwell last century and usually referred to as the paradox of Maxwell's demon. Consider a sealed box divided into two by a membrane equipped with a shutter. The box is filled on both sides of the membrane by a gas at a uniform temperature and pressure. Being in thermodynamic equilibrium, the system is at a state of maximum entropy with no reserves of useful energy to do work. Nothing further of interest will happen, save for the random cavorting of the gas molecules.

Suppose, however, that there is a tiny demon inside the box who can operate the shutter mechanism. He notices that the motions of the molecules, being chaotic, involve a whole range of speeds and directions. Some molecules move rapidly, others slowly. The *average* speed is the factor that determines the temperature of the gas. That does not change. But individual molecules change speed and direction every time they suffer collisions with their neighbours, or with the walls of the box. The demon then adopts the following strategy. Whenever a fast molecule approaches from the right hand half of the container, he opens the shutter and lets the molecule pass through into the left-hand chamber. Conversely, slow molecules that approach from the other direction are admitted into the right-hand chamber. After a while, the left-hand chamber will be full of rather rapidly moving molecules (on average), while the right-hand chamber will be full of slower molecules. The left-hand chamber will therefore be at a higher temperature than the right. The demon, by dextrous and prompt manipulation of individual molecules, will have created a temperature difference between the two chambers. Equilibrium will no longer prevail, and so the entropy will have been reduced. It would now be possible to use the temperature difference to execute some useful work (by driving a heat engine for instance) until the useful energy was once again dissipated and equilibrium restored. The demon could then repeat his act, and we should have at our disposal the basis of a perpetuum mobile. By operating this type of sorting strategy at the cosmic level an omnipresent demon could prevent the universe from descending into the heat death.

Alas, closer inspection proves that Maxwell's demon will not work. In the 1920s Leo Szilard investigated the demon's operation in greater detail. He realized that in order to operate successfully the demon must have precise information about the speed of the approaching molecules. This information can only be achieved at a price, and the price paid is in the thermodynamic currency of entropy. For example, an approaching molecule could be illuminated by a strong flux of light, and the Doppler effect used to measure its speed in the same way that police use car radar traps. But the expenditure of useful energy entailed in this step would raise the entropy of the gas by more than it would be lowered as a result of the sorting effect. Evidently the second law cannot be defeated even by intelligent manipulation at the molecular level.

If these thermodynamic ideas are correct, no natural agency, intelligent or otherwise, can delay for ever the end of the universe. As we have seen, if the universe continues to expand, it may never reach precise thermodynamic equilibrium. Nevertheless, the organization that we now perceive is inevitably destined to decline to a level where the cosmos would bear no resemblance to its present lively phase. Only a supernatural God could truly wind it up again.

16. Is the universe a 'free lunch'?

'Nothing can be created out of nothing.'

Lucretius

We can now draw together all the strands of our investigation to construct a cosmic scenario that reveals the astonishing scope of the new physics to explain the physical world. I am not suggesting the scenario should be taken too seriously (though it is being discussed seriously by physicists). It does, however, illustrate the sort of ideas that modern physics has thrown up — ideas which cannot be ignored in our search for God.

In the Preface I posed what I called the challenge of the Big Four questions of existence: 'Why are the laws of nature what they are? Why does the universe consist of the things it does? How did those things arise? How did the universe achieve its organization?

The new physics has gone a long way to provide answers for these questions. To take them in reverse order, we have seen how an initially chaotic state can evolve into a more orderly one provided there is a supply of negative entropy. We have also seen how that negative entropy can be generated by the expansion of the universe, so that there is no longer any need to assume, as did the scientists of earlier generations, that the universe was somehow created in a highly organized, specially arranged state. The present organization is consistent with a universe that began accidentally in a random state.

The question of the origin of physical things has been discussed in detail in the early chapters. Objects such as stars and planets are known to have formed from the primeval gases, while the cosmic material itself was evidently created in the big bang. Recent discoveries in particle physics have suggested mechanisms whereby matter can be created in empty space by the cosmic gravitational field, which only leaves the origin of spacetime itself as a mystery. But even here there are some indications that space and time could have sprung into existence spontaneously without violating the laws of physics. The reason for this bizarre possibility concerns the quantum theory.

We have seen how the quantum factor permits events to occur without causes in the subatomic world. Particles, for example, can appear out of nowhere without specific causation. When the quantum theory is extended to gravity, it involves the behaviour of spacetime itself. Although there is still no satisfactory theory of quantum gravity, physicists have a good idea of the broad features that would be entailed in such a theory. It would, for example, endow space and time with the same sort of fuzzy unpredictability that characterizes quantum matter. In particular, it would allow spacetime to be created and destroyed spontaneously and uncaused in the same way that particles are created and destroyed spontaneously and uncaused. The theory would entail a certain mathematically determined probability that, for instance, a blob of space would appear where none existed before. Thus, spacetime could pop out of nothingness as the result of a causeless quantum transition.

On general grounds the abrupt appearance of spacetime by a quantum mechanism might be expected to occur solely on an ultramicroscopic scale, because quantum processes usually apply only to microscopic phenomena. Indeed, the spontaneously created space might typically be only 10^{-33} cm in size. This finite blob of space need have no edges, however: it could be closed into a hypersphere as described in Chapter 2. Probably such a mini-universe would rapidly disappear again by another, reverse, quantum fluctuation. Nevertheless, there is a chance that, rather than fading away, the newly created blob of space will suddenly begin to inflate like a balloon.

The origin of this behaviour would lie with other quantum processes associated, not with gravity, but with the remaining forces of nature. In Chapter 13 I briefly described the so-called 'inflationary universe scenario' in which the 'grand unified force' causes the nascent universe to become unstable and embark upon a phase of runaway exponential expansion. In this way the quantum microworld could swell to cosmic proportions in a minute fraction of a second. The energy accumulated in this big bang would, at the abrupt termination of the inflationary phase, became converted into matter and radiation, and the universe would then proceed according to conventional understanding.

In this remarkable scenario, the entire cosmos simply comes out of nowhere, completely in accordance with the laws of quantum physics, and creates along the way all the matter and energy needed to build the universe we now see. It thus incorporates the creation of all physical things, including space and time. Rather than postulate an unknowable singularity to start the universe off (see Chapter 2), the quantum spacetime model attempts to explain everything entirely within the context of the laws of physics. It is an awesome claim. We are used to the idea of 'putting something in and getting something out', but getting something for nothing (or out of nothing) is alien. Yet the world of quantum physics routinely produces something for nothing. Quantum gravity suggests we might get everything for nothing. Discussing this scenario, the physicist Alan Guth remarked: 'It is often said that there is no such thing as a free lunch. The universe, however, is a free lunch.'[1]

Does such a universe model have any need for God? In Chapter 3 we saw how one traditional cosmological argument for God proceeded on the assumption that everything must have a cause. Quantum physics has confounded this claim. But what of the remaining two questions? Why does the universe possess the things and the laws it does? Can science provide an answer?

In Chapter 11 it was explained how the goal of the so-called supergravity theory is to provide a mathematical description for all the forces of nature and all the fundamental particles of matter. If successful, this theory would reduce the remaining two questions to one. The 'things' of which the world is composed — protons, neutrons, mesons, electrons, and so on — would be accounted for within the framework of the supergravity theory. At present, the status of the physical laws is rather different. We generally know how an electron or a proton behaves once we have it, but we have no real idea of why there *are* electrons or protons and not particles of quite different properties. If supergravity is fully successful, it will tell us not only why there are the particles that exist, but also why they have the masses, charges and other properties that they do.

All of this will follow from a magnificent mathematical theory that will encompass all of physics (in the reductionist sense) in one superlaw. But we come back still to the question: why *that* superlaw?

We have thus reached the ultimate question of existence. Physics can perhaps explain the content, origin and organization of the physical universe, but not the laws (or superlaw) of physics itself. Traditionally, God is credited with having invented the laws of nature and created the things (spacetime, atoms, people, among others) on which those laws operate. The 'free lunch' scenario claims that all you need are the laws — the universe can take care of itself, including its own creation.

But what of the laws? They have to be 'there' to start with so that the universe can come into being. Quantum physics has to exist (in some sense) so that a quantum transition can generate the cosmos in the first place. Many scientists believe that the question of why the laws of physics are what they are is meaningless, or at least cannot be answered scientifically. Others have argued 'anthropically', maintaining that the laws must be such as to admit observers. But there is a further possibility. Perhaps the laws — or the ultimate superlaw — will emerge to be the only *logically* possible physical principle. We turn to this idea in the final chapter.

17. The physicist's conception of nature

'Nature has a simplicity and therefore a great beauty'
Richard Feynman

'If you get simple beauty and nought else,
You get about the best thing God invents'
Elizabeth Browning

In the foregoing chapters we have explored the implications for religion of recent advances in science, particularly in what has become known as the new physics. In spite of the spectacular success of modern science, it would be foolish to suppose that the fundamental questions concerning the existence of God, the purpose of the universe or the role of mankind in the natural and supernatural scheme has been answered by these advances. Indeed, scientists themselves have a wide range of religious beliefs.

It is often claimed that science and religion can peacefully coexist because they address themselves to different issues. Questions of religious doctrine, such as morality or the concept of the trinity, are intrinsically different from scientific questions, like deciding the best mathematical description of gravity. Yet it cannot be denied that science does have something to say about religious matters. In topics such as the nature of time, the origin of matter and life, or causality and determinism, the very conceptual framework in which the religious questions are posed can be altered by scientific advances. Some of the major theological issues of several centuries ago (such as the location of Heaven and Hell) have been rendered meaningless by modern cosmology and our improved understanding of the nature of space and time.

Many people are inclined to cast the conflict between science and religion in terms of 'right and wrong'. It is tempting to believe that there exists an ultimate truth — an objective reality — that both science and religion alike are groping towards. According to this eminently reasonable position questions such as 'Does God exist?', 'Are there supernatural miracles?', 'Was there a creation?', 'Is there a purpose in the universe?', 'Did life arise by accident?', all have answers 'Yes' or 'No' even though we may not know them.

One frequently encounters the view that scientific theories are approximations to an actual reality. As our understanding advances, so the fit between theory and reality improves. According to this perspective 'the true' laws of nature are buried in the data of observation and experiment, to be dug out by persistent and inspired investigation. One day, so the philosophy goes, we can expect to expose the *correct* laws, to which our present textbook laws are a creditable but flawed facsimile. In many ways, this is the aim of the supergravity programme, whose proponents anticipate the discovery of a set of equations which will embody 'the true' laws in their entirety.

Not all physicists, however, believe that it is meaningful to talk about 'truth'. Physics, according to this alternative philosophy, is not about truth at all, but about models: models that help us to relate one observation to another in a systematic way. Niels Bohr expressed this so-called positivist view when he said that physics tells what we can *know* about the universe, not how it *is*. As explained in Chapter 8, the quantum theory has prompted many physicists to declare that there is no 'objective' reality at all. The only reality is that revealed through our observations. Adopting this view, it is not possible to pronounce a particular theory 'right' or 'wrong', merely that it is useful or less useful, a useful theory being one which connects a wide range of phenomena in a single descriptive scheme to high accuracy. Such a view, then, is diametrically opposite to that of religion, in which the adherent believes in an ultimate truth. A religious proposition is usually regarded as either right or wrong, not as some sort of model of our experiences.

The difference in approach is illustrated by the willingness of physicists to abandon a particular cherished theory in favour of a better one. As Robert Merton once wrote: 'Most institutions demand unqualified faith, but the institution of science makes scepticism a virtue.' When Einstein discovered the theory of relativity, it was realized that Newton's theory of space, time and mechanics was inadequate for describing the behaviour of bodies moving close to the speed of light, and so it was replaced. Newton's theory is not really wrong, it merely has a limited range of validity. The special theory of relativity is a more useful theory (which reduces to Newton's theory for the case when speeds are low) but which gives a more accurate account of high-speed systems. This theory has in turn been superseded by the so-called general theory of relativity, and few physicists doubt that, in turn, the general theory will be improved upon. As regards a 'final', perfect theory that cannot be improved upon, some physicists regard such a thing to be as meaningless as the idea of a perfect picture or a perfect symphony.

The ability of the scientific method to accommodate change in the light of new discoveries represents one of science's great strengths. By basing itself on utility rather than truth, science distinguishes itself sharply from religion. Religion is founded on dogma and received wisdom, which purports to represent immutable truth. Though peripheral doctrinal issues may become adapted and distorted with time, the idea of a religion's fundamental dogma being abandoned in favour of a more accurate 'model' of reality is unthinkable. If the Church pronounced that, on the basis of new evidence, it appeared that Christ was not resurrected after all, then Christianity could hardly survive in any recognizable form. Some critics have claimed that dogmatic rigidity means that every new discovery and every novel idea is likely to pose a threat to religion, whereas new facts and ideas are the very life-blood of science. So it is that scientific discoveries have, over the years, set science and religion into conflict.

In spite of the fact that religion looks backward to revealed truth while science looks forward to new vistas and discoveries, both activities produce a sense of awe and a curious mixture of humility and arrogance in their practitioners. All great scientists are inspired by the subtlety and beauty of the natural world that they are seeking to understand. Each new subatomic particle, every unexpected astronomical object, produces delight and wonderment. In constructing their theories, physicists are frequently guided by arcane concepts of elegance in the belief that the universe is intrinsically beautiful. Time and again this artistic taste has proved a fruitful guiding principle and led directly to new discoveries, even when it at first sight appears to contradict the observational facts.

Paul Dirac once wrote:

> It is more important to have beauty in one's equations than to

> have them fit experiment . . . because the discrepancy may be due to minor features that are not properly taken into account and that will get cleared up with further developments of the theory . . . It seems that if one is working from the point of view of getting beauty in one's equations, and if one has really a sound insight, one is on a sure line of progress.[1]

It is an idea succinctly captured by Bohm: 'Physics is a form of insight and as such it's a form of art.'[2]

Einstein, while discussing his distrust of the idea of a personal God, nevertheless expressed his admiration for 'the beauty of . . . the logical simplicity of the order and harmony which we can grasp humbly and only imperfectly.'

Central to the physicist's notion of beauty are *harmony*, *simplicity* and *symmetry*. Einstein again:

> All of these endeavours are based on the belief that existence should have a completely harmonious structure. Today we have less ground than ever before for allowing ourselves to be forced away from this wonderful belief. Equations of such complexity as are the equations of the gravitational field can be found only through the discovery of a logically simple mathematical condition.[3]

In more recent times this sentiment has been echoed by Wheeler:

> The beauty in the laws of physics is the fantastic simplicity that they have . . . What is the ultimate mathematical machinery behind it all? That's surely the most beautiful of all.[4]

Today this guiding principle motivates the search for the superforce. In a review of progress in the mathematics of supergravity, two leading exponents recently remarked: 'In the derivation of all forces from the common requirement of local symmetry one can glimpse a deeply satisfying order.'[5]

When physicists talk of beauty and symmetry the language through which these concepts are expressed is mathematics. It cannot be overstressed how important mathematics is to science in general and physics in particular. Leonardo da Vinci once wrote: 'No human investigation can be called real science if it cannot be demonstrated mathematically.' This is probably more true today than it was in the fifteenth century.

The neurotic fear of mathematics experienced by most ordinary people is chiefly responsible for their estrangement with physical science. It is a barrier that effectively cuts them off from a full appreciation of scientific discoveries, and prevents them from enjoying vast areas of nature that have been revealed through painstaking research. For, as Roger Bacon appreciated: 'Mathematics is the door and the key to the sciences . . . For the things of this world cannot be made known without a knowledge of mathematics.'[6]

Many physicists have become so deeply impressed with the mathematical simplicity and elegance of the laws of nature that they maintain it reveals a fundamental feature of existence. Sir James Jeans once remarked that in his opinion 'God is a mathematician'. But why should God choose to implement his ideas in mathematical form?

Mathematics is the poetry of logic. No expression of law can be more compelling or more satisfying than one based on simple and unassailable logical foundations. In the words of John Wheeler:

> Little astonishment should there be, therefore, if the description of nature carries one in the end to logic, the ethereal eyrie at the center of mathematics. If, as one believes, all mathematics reduces to the mathematics of logic, and all physics reduces to mathematics, what alternative is there but for all physics to reduce to the mathematics of logic? Logic is the only branch of mathematics that can 'think about itself'.[7]

One of the attractions of the logical expression of nature is the possibility that much, if not all of nature, might be deduced from logical inference rather than empirical evidence. Before the Second World War, Arthur Eddington and E.A. Milne both attempted to construct deductive theories of the universe (without much success). The idea raises an intriguing prospect. Could it be that the universe is the way it is because it is an *inevitable* consequence of logical necessity? The great French scientist Jean d'Alembert wrote: 'To someone who could grasp the universe from a unified standpoint, the entire creation would appear as a unique truth and necessity.' It is an idea that casts a curious light on the issue of God's omnipotence. In Chapter 10 it was pointed out that an omnipotent creator could fashion any universe he desired. Christians claim that this *particular* universe can be explained as God's choice, taken from an infinite range of alternatives, for reasons that are unknown to us. But even an omnipotent God cannot break the rules of logic. God cannot make 2 = 3, or make a square a circle. The hasty assumption that God can create *any* universe must be qualified by the restriction that it be logically consistent. Now if there exists only *one* logically consistent universe then God would effectively have had no choice at all. Einstein noted: 'What I'm really interested in is whether God could have made the world in a different way; that is, whether the necessity of logical simplicity leaves any freedom at all.'[8]

If there really is only one possible sort of creation, why do we need a creator at all? What function could he have save for 'pushing the button' to set the thing going? But such a function does not require a *mind* — it would merely be a triggering mechanism and, as we saw in the previous chapter, even that is not needed in the world of quantum physics. So does this philosophy of a unique physical solution to the fundamental logical–mathematical equation of the universe deny the existence of God? Indeed not. It makes redundant the idea of God-the-creator, but it does not rule out a universal mind existing as part of that unique physical universe: a natural, as opposed to supernatural God. Of course 'part of' in this context does not mean 'located somewhere in space' any more than our own minds can be located in space. Nor does it mean 'made out of atoms' any more than our minds (as opposed to brains) are made out of atoms. The brain is the medium of expression of the human mind. Similarly the entire physical universe would be the medium of expression of the mind of a natural God. In this context, God is the supreme holistic concept, perhaps many levels of description above that of the human mind.

If such ideas are accepted it becomes of crucial importance to know the origin and fate of the physical universe. Because mind requires organization, its existence is threatened by the second law of thermodynamics. As the universe is slowly choked to death by its own entropy, will God die too? The alternative — gravitational collapse to a singularity resulting in the total obliteration of the physical universe — seems even less promising. Only a universe of the cyclic or steady-state varieties would appear to offer scope for a natural God to be both infinite and eternal.

So far our discussion of the physicist's conception of nature has dwelt upon the reductionist approach. The sense of beauty and simplicity that so inspires the physicist in his search for new laws and models largely refers to the elementary structures that go to build up the world: the subatomic particles such as quarks and leptons, and the fundamental forces that operate between them. But the holistic aspect of God reminds us again that however well the physicist may come to understand what the world is made of and how it is put together, the holistic features will not be encompassed by any purely reductionist conception.

Richard Feynman once put it this way:

> We have a way of discussing the world, when we talk of it at various hierarchies, or levels. Now I do not mean to be very precise, dividing the world into definite levels, but I will indicate,

by describing a set of ideas, what I mean by hierarchies of ideas.

For example, at one end we have the fundamental laws of physics. Then we invent other terms for concepts which are approximate, which have, we believe, their ultimate explanation in terms of the fundamental laws. For instance, 'heat'. Heat is supposed to be jiggling, and the word for a hot thing is just the word for a mass of atoms which are jiggling. But for a while, if we are talking about heat, we sometimes forget about the atoms jiggling — just as when we talk about the glacier we do not always think of the hexagonal ice and the snowflakes which originally fell. Another example of the same thing is a salt crystal. Looked at fundamentally it is a lot of protons, neutrons, and electrons; but we have this concept 'salt crystal', which carries a whole pattern already of fundamental interactions. An idea like pressure is the same.

Now if we go higher up from this, in another level we have properties of substances — like 'refractive index', how light is bent when it goes through something; or 'surface tension', the fact that water tends to pull itself together, both of which are described by numbers. I remind you that we have to go through several laws down to find out that it is the pull of the atoms, and so on. But we still say 'surface tension', and do not always worry, when discussing surface tension, about the inner workings.

On, up in the hierarchy. With the water we have waves, and we have a thing like a storm, the word 'storm' which represents an enormous mass of phenomena, or a 'sun spot', or 'star', which is an accumulation of things. And it is not worth while always to think of it way back. In fact we cannot, because the higher up we go the more steps we have in between, each one of which is a little weak. We have not thought them all through yet.

As we go up in this hierarchy of complexity, we get to things like muscle twitch, or nerve impulse, which is an enormously complicated thing in the physical world, involving an organization of matter in a very elaborate complexity. Then come things like 'frog'.

And then we go on, and we come to words and concepts like 'man' and 'history', or 'political expediency', and so forth, a series of concepts which we use to understand things at an ever higher level.

And going on, we come to things like evil, and beauty, and hope . .

> Which end is nearer to God; if I may use a religious metaphor. Beauty and hope, or the fundamental laws? I think that the right way, of course, is to say that what we have to look at is the whole structural interconnection of the thing; and that all the sciences, and not just the sciences but all the efforts of intellectual kinds, are an endeavour to see the connections of the hierarchies, to connect beauty to history, to connect history to man's psychology, man's psychology to the working of the brain, the brain to the neural impulse, the neural impulse to the chemistry, and so forth, up and down, both ways. And today we cannot, and it is no use making believe that we can, draw carefully a line all the way from one end of this thing to the other, because we have only just begun to see that there is this relative hierarchy.
>
> And I do not think either end is nearer to God.[9]

As emphasized in the earlier chapters there is a growing appreciation among scientists of the importance of the structural hierarchy in nature: that holistic concepts like life, organization and mind are indeed meaningful, and they cannot be explained away as 'nothing but' atoms or quarks or unified forces or whatever. However important it may be to understand the fundamental simplicity at the heart of all natural phenomena, it cannot be the whole story. The complexity is just as important.

One of the major unsolved problems of modern physics is whether the holistic features of a physical system require additional holistic laws that cannot be reduced to the fundamental laws of elementary forces and particles. So far we have no evidence for truly holistic laws of physics. Thermodynamics, for example, treats holistic systems such as gases, containing enormous numbers of molecules which act collectively. Concepts such as temperature and pressure are meaningless at the level of individual molecules. Yet the gas laws can all be derived from the lower-level laws of molecular motion applied in a statistical way to large collections. A truly holistic law would be, for example, a case where a new force or organizing influence emerged at the collective level that did not have its origin in the component parts individually. This was the assumption of vitalism in the explanation of life.

A more striking example of a holistic law of physics would be psychokinesis or telepathy. Proponents of so-called paranormal phenomena claim that the human mind can actually exert forces on distant matter. Presumably such forces are unknown at the reductionist level: they are not nuclear, gravitational or electromagnetic. The most direct illustration of these psychic forces is in the spectacular cases of remote metal bending, where the subject appears to deform a metallic object by mind-power alone, without physical contact. The author has devised an extremely stringent test of this phenomenon using metal rods sealed inside glass containers from which the air has been replaced by a secret combination of rare gases to preclude tampering. In a recent trial of arch-metal benders not one was able to produce any measurable deformation.

In the foregoing it has been suggested that perhaps the structure of the physical world results in part or in whole from the operation of exceedingly simple logical principles expressed in elementary mathematical form. One difficulty in accepting this idea is the problem of complexity. Can we really believe that, for example, life and mind arise solely from logical rules, rather than holistic forces?

There is a beautiful illustration that interesting and complex activity can indeed be generated by the operation of the simplest imaginable logical rules. The Cambridge mathematician John Conway has invented a scheme known as Life, which is a simple game played by a single player on a board marked into many squares (cells). Black counters are placed in some of the squares and the resulting configuration of counters changes shape according to a set of rules:

1. Every counter with 2 or 3 neighbouring counters survives to the next generation (i.e. the next move).
2. Every counter with 0 to 1 neighbours 'dies' (of loneliness) and every counter with 4 or more neighbours dies (of overcrowding).
3. Every empty cell with exactly 3 neighbouring occupied cells gives birth to a new counter.

With these simple rules of birth, survival and death, Conway and his colleagues have discovered the most astonishing richness and variety in the evolution of certain counter configurations. Two features in particular are striking. The first is that simple shapes can evolve into

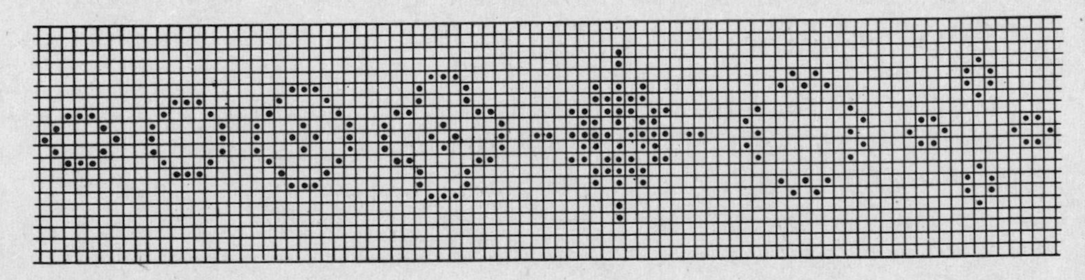

25 In John Conway's game *Life*, the above sequence of patterns evolves (some intermediate steps have been omitted). The shapes are coincidentally reminiscent of the life-cycle of a flower.

complex structures. Consider, for example, the 'seed' shown in Fig. 25 which grows into a flower, withers, and dies, leaving four smaller 'seeds'.

Still more remarkable is the discovery that certain shapes retain a coherence, displaying activity that is reminiscent of *behaviour*. The simplest example is the 'glider' which holds together and moves across the board (see Fig. 26). Larger groupings known as 'spaceships' leave a trail of 'sparks' as they move. Much larger 'spaceships', however, require a collection of smaller 'escorts' to eat up debris that the large spaceship ejects ahead of it and which would otherwise block its path, causing it to break up.

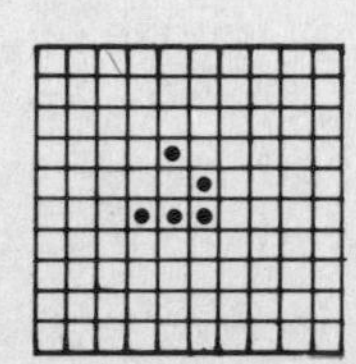

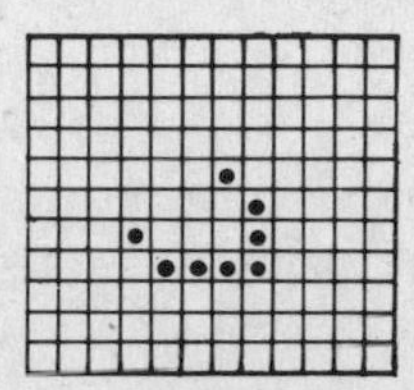

26 The simple arrangement of five dots, known as a 'glider', has the curious property that it travels diagonally across the board, but remains intact. The larger collection of eight dots is called a 'spaceship' and also travels intact, but emits 'sparks' as it goes.

Conway's game, with the assistance of computers, can be used to test conjectures about self-reproducing machines and other abstract logico-mathematical puzzles. This involves constructing shapes that can generate other shapes on a production line basis. One example is a 'glider gun' that produces a new glider every thirty moves. Such a device can be constructed from the debris of a thirteen glider collision! By careful positioning of glider guns, the intersecting beams of gliders build a factory which launches a spaceship every 300 moves. All of this 'behaviour' is *automatic*. Given the chosen initial configuration, the game itself generates the structures and activity: there is no human input. And all this from a few simple logical rules.

Physics, in my opinion, makes its chief contribution through reductionism. The holistic aspects fall more appropriately into the cognitive sciences and subjects like systems theory, games theory, sociology and politics. That is not to claim that physics has nothing to say about holism, for clearly it does. Thermodynamics, the quantum theory and the physics of self-organizing systems all involve holistic concepts. Nevertheless I don't believe that physics can tackle questions about, for example, purpose or morality.

I am sometimes asked whether the insight which physicists have gained into the inner workings of nature through the study of fundamental processes throws any light on the nature of God's plan for the universe, or reveals the struggle between good and evil. It does not. There is nothing good or evil about the way quarks are united into protons and neutrons, or the absorption and emission of quanta, the bending of spacetime by matter, the abstract symmetries that unite the fundamental particles, and so on. True, there is much competition to be found in nature, between the balance and interplay of different forces, for instance. A star is a battleground of opposing forces. Gravity, which tries to crush the star, struggles against the forces of thermal pressure and electromagnetic radiation which try to explode it — forces which in turn are generated by the release of energy due to nuclear interactions. And all across the universe the struggle goes on. However, if the opposing forces were not more or less equally matched, all physical systems would be overwhelmed by one or the other, and activity would soon cease. The universe is complex and interesting precisely because these battles continue over the aeons.

The great drama unfolds in the oceans of time provided by this cosmic stalemate. There is some puzzle over these apparently accidental 'hang-ups', as Freeman Dyson calls them:

> Since the universe is on a one-way slide toward a state of final death in which energy is maximally degraded, how does it manage, like King Charles, to take such an unconscionably long time a-dying?[10]

This fortuitous stability, wherein the universe postpones its descent into total chaos for an astronomical duration, is another aspect of the 'coincidences' discussed in Chapter 13.

There is the size hang-up, which prevents the sudden collapse of the universe under its own gravity; the free-fall time for implosion to a big crunch (if it occurs) is many billions of years, due to the exceedingly spacious distribution of cosmic matter. Then there is the spin hang-up that stabilizes galaxies and planetary systems from simply falling in on themselves. Centrifugal force comes to the rescue to counterbalance the inward pull of gravity. Finally, there are nuclear hang-ups which ensure that the rate of nuclear fuel consumption in stars is very gradual.

These hang-ups do not last for ever. When they fail, violence frequently erupts. The universe is full of violent activity: the explosion of stars, huge eruptions of energy from disturbed galaxies and quasars, horrendous collisions between monstrous objects, bodies torn apart by gravity, matter crushed to oblivion in black holes. Awesome violence. Yet the physicist sees nothing evil in this violence as such. Amid the turmoil of unleashed energy, nature may sow the seeds of future tranquility. The heavy elements that make up our equable planet were created in the fire and explosion of supernovae long ago. The entire universe was born in an outburst of incomparable, unlimited violence. To the physicist, violent phenomena are simply one particular expression of natural laws that are morally neutral. Good and evil apply only to mind, not matter.

In the foregoing chapters we have searched across the range of modern physics — the new ideas about space and time, order and disorder, mind and matter — in our search for God. Much of what has been presented will no doubt confirm the opinion of some that science is implacably opposed to religion and continues to threaten the very basis of most religious doctrine. It would be foolish to deny that many of the traditional religious ideas about God, man and the nature of the universe have been swept away by the new physics. But our search has turned up many positive signs too. The existence of mind, for example, as an abstract, holistic, organizational pattern, capable even of disembodiment, refutes the reductionist philosophy that we are all nothing but moving mounds of atoms.

However, it was not my intention in this book to provide easy answers for longstanding religious questions. What I have sought to do is to expand the context in which the traditional religious issues are discussed. The new physics has overturned so many commonsense notions of space, time and matter that no serious religious thinker can ignore it.

I began by making the claim that science offers a surer path than religion in the search of God. It is my deep conviction that only by understanding the world in all its many aspects — reductionist and holist, mathematical and poetical, through forces, fields, and particles as well as through good and evil — that we will come to understand ourselves and the meaning behind this universe, our home.

Footnotes

Chapter 1

1 H. Bondi, 'Religion is a good thing', *Lying Truths* (eds. R. Duncan and M. Weston-Smith; Pergamon 1979).

2 Ibid.

3 K. Pedlar, *Mind over Matter* (Thames Methuen 1981), p.11.

4 H. Morowitz, 'Rediscovering the mind', *The Mind's I* (eds. D.R. Hofstadter and D.C. Dennett; Harvester/Basic Books 1981).

Chapter 2

1 I. Kant, *Critique of Pure Reason* (ed. J.M.D. Meiklejohn; Dent 1934, 1945; first published 1781).

2 The main part of Pope Pius XII's address is reprinted in English in *Bulletin of the Atomic Scientists 8*, pp.143–6, 165 (1952).

3 E. McMullin, 'How should cosmology relate to theology?', *The Sciences and Theology in the Twentieth Century* (ed. A.R. Peacocke; Oriel 1981).

Chapter 3

1 Samuel Clarke developed his version of the cosmological argument in the Boyle lectures delivered in 1704, subsequently published under the title *A Demonstration of the Being and Attributes of God*. Together with a further series of lectures, given in 1905, the lectures were reprinted as *A Discourse Concerning the Being and Attributes of God, the Obligations of Natural Religion, and the Truth and Certainty of the Christian Revelation* (John and Paul Knapton; London, 1738, ninth edition).

2 Aquinas, *Summa Theologiae* (ed. T. Gilby; Eyre & Spottiswoode 1964).

3 Clark, op. cit. pp.12–13.
4 D. Hume, *Dialogues Concerning Natural Religion* (ed. H.D. Aiken; Hafner 1969; first published 1779). Part IV.
5 St. Augustine of Hippo, 'On the beginning of time', *The City of God* (trans. M. Dods; Hafner 1948).
6 J.A. Wheeler, 'Genesis and observership', *Foundational Problems in the Special Sciences* (eds. R.E. Butts and K.J. Hintikka; Reidel 1977).
7 Ibid.
8 J.A. Wheeler, 'Beyond the black hole', *Some Strangeness in the Proportion* (ed. H. Woolf; Addison-Wesley 1980). See also 'Is physics legislated by cosmology', *Quantum Gravity: An Oxford Symposium* (eds. C.J. Isham, R. Penrose and D.W. Sciama; Oxford: Clarendon Press 1975) and *Frontiers of Time* (North-Holland 1979) by the same author.
9 A so-called 'bubble cosmology' has been proposed by J.R. Gott III in *Nature 295*, 304 (1982), and described less formally in *The Sciences* (1982). Similar ideas have been published by Katsuhiko Sato et. al. in *Progress in Theoretical Physics* (Letters) *65*, 1443 (1981).
10 R. Swinburne, *The Existence of God* (Oxford: Clarendon Press 1979) p.122.

Chapter 4

1 Swinburne, op. cit., chapter 7.
2 Ibid. pp.131–2.
3 E.W. Barnes, *Scientific Theory and Religion* (Cambridge University Press 1933) p.595.
4 P.C.W. Davies, *The Physics of Time Asymmetry* (Surrey University Press/University of California Press 1974).
5 S.W. Hawking, 'Breakdown of predictability in gravitational collapse', *Physical Review D 14*, 2460 (1976); see also *Scientific American 236*, 34 (1977).
6 Ibid.

Chapter 5

1 A. Koestler, 'Nothing but . . .?', *Lying Truths* (eds. R. Duncan and M. Weston-Smith; Pergamon 1979). See also *Beyond Reductionism — New Perspectives in the Life Sciences* (eds. A. Koestler and J.R. Smythies; Hutchinson 1969), and *Janus: A Summing Up* (Koestler; Vintage 1979).

2 D. Bohm, *Wholeness and the Implicate Order* (Routledge & Kegan Paul 1980).

3 E. Schrödinger, *What is Life?* (Cambridge University Press 1946) p.77. This book is now reprinted together with Schrödinger's other classic *Mind and Matter* (1958), under the joint titles by CUP (1967).

4 Ibid., p.76.

5 F. Crick, *Life Itself: Its Origin and Nature* (Macdonald/Simon & Schuster 1982).

6 E. McMullin, op. cit., p.47.

Chapter 6

1 *New Catholic Encyclopedia* (McGraw-Hill 1967) vol. 13, p.460.

2 R.J. Hirst, *The Problems of Perception* (Allen & Unwin 1959) p.181. Descartes' paradigm for dualism was set out in his main works *Le Discours de la Methode* (1637), *Meditationes de Prima Philosophia* (1641) and *Principia Philosophiae* (1644). See *The Philosophical Works of Descartes* (trans. E.S. Haldane and G.R.J. Ross; 2 vols., Cambridge University Press 1967).

3 G. Ryle, *The Concept of Mind* (Hutchinson 1949, many reprints).

4 Ibid., p.20.

5 *New Catholic Encyclopedia*, op. cit., p.471.

6 Ryle, op. cit., p.23.

7 *The Mind's I*, p.6.

8 D.R. Hofstadter, *Gödel, Escher, Bach* (Basic Books, 1979) p.577.

9 D.M. MacKay, *The Clockwork Image* (Inter-Varsity Press, 1974) chapter 9.

10 J.A. Fodor, 'The mind-body problem', *Scientific American* (January 1981).

Chapter 7

1 T. Reid, *Essays on the Intellectual Powers of Man* (ed. A.D. Woozley; MIT Press 1969; first published 1785) Essay III, chapter 4.

2 Ryle, op. cit., p.187.

3 D. Hume, *A Treatise of Human Nature* (ed. P.H. Nidditch; Oxford University Press 1978; first published 1739) Book I, Part 4, chapter 6.

4 J. Locke, *Essay Concerning Human Understanding*, chapter 27 (ed. A.D. Woozley; Dent 1976; first published 1690).

5 J.R. Lucas, 'Minds, machines and Gödel', *Minds and Machines* (ed. A.R. Anderson; Prentice-Hall 1964) p.57.
6 A.J. Ayer, *The Central Questions of Philosophy* (Weidenfeld & Nicolson 1973; Penguin 1977) p.119.
7 Hofstadter, op. cit., p.697.
8 Lucas, op. cit.
9 Hofstadter, op. cit., p.709.
10 See 'Where am I?', *Brainstorms* by D.C. Dennett (Bradford Books 1978).
11 MacKay, op. cit., p.75.

Chapter 8

1 D. Bohm, op. cit., p.134.
2 E. Wigner, 'Remarks on the mind-body question', *The Scientist Speculates* (ed. I.J. Good; Heinemann 1962).
3 B.S. DeWitt, 'The many-universes interpretation of quantum mechanics', *Foundations of Quantum Mechanics* (ed. B. d'Espagnat; Academic Press 1971).

Chapter 9

1 See, for example, 'McTaggart, fixity and coming true', *Time, Reduction and Reality* (ed. Richard Healey; Cambridge University Press 1981).
2 St. Anselm, *Proslogion* (trans. M.J. Charlesworth; University of Notre Dame Press 1979) chapter XIX. One of the earliest proponents of God's timelessness was A.M.S. Boethius (c. 480–524). See *The Consolation of Philosophy* (ed. W. Anderson; Centaur 1963) section 5.6.
3 P. Tillich, *Systematic Theology* (S.C.M. 1978) vol. I, p.305.
4 K. Barth, *Church Dogmatics* II (i) (trans. G.W. Bromiley and T.F. Torrance; T. & T. Clark 1956) p.620.

Chapter 10

1 MacKay, op. cit., p.78.

Chapter 11

1 J.A. Wheeler in *Gravitation* (eds. C.W. Misner, K.S. Thorne and J.A. Wheeler; Freeman 1973) p.1197.

Chapter 12

1 *The Works of William Paley* (Oxford: Clarendon Press 1938) vol. IV, p.1.

2 R. Swinburne, *The Existence of God* (Oxford: Clarendon Press 1979) chapter 8.

3 B. Carter, 'Large number coincidences and the anthropic principle in cosmology', *Confrontation of Cosmological Theories with Observation* (ed. M.S. Longair; Reidel 1974). This paper is based on a more extensive earlier study by Carter entitled 'The significance of numerical coincidences in nature', which has, however, never been published in its entirety.

Chapter 13

1 R. Penrose, 'Singularities and time-asymmetry', *General Relativity: An Einstein Centenary Survey* (eds. S.W. Hawking and W. Israel; Cambridge University Press 1979).

2 Carter, op. cit.

Chapter 15

1 E.R. Harrison, *Cosmology* (Cambridge University Press 1981) p.360.

2 N. Nicholson, 'The expanding universe' in *The Pot Geranium* (Faber & Faber 1954).

Chapter 16

1 A. H. Guth, 'Speculations on the origin of the matter, energy and entropy of the universe' in *Asymptotic Realms of Physics: A Festschrift in Honor of Francis Low* (eds. A.H. Guth, K. Huang and R.L. Jaffe; MIT Press 1983).

Chapter 17

1 P.A.M. Dirac, 'The evolution of the physicist's picture of nature', *Scientific American* (May 1963).

2 D. Bohm, in *A Question of Physics: Conversations in Physics and Biology* (ed. P. Buckley and F.D. Peat; Routledge & Kegan Paul 1979) p.129.

3 A. Einstein, *Essays in Science* (Philosophical Library, New York 1934).

4 J.A. Wheeler in Buckley and Peat, op. cit., p.60.

5 D.Z. Freedman and P. van Nieuwenhuizen, 'Supergravity and the unification of the laws of physics', *Scientific American* (February 1978).

6 R. Bacon, *Opus Majus* (trans. Robert Belle Burke; University of Pennsylvania Press 1928).

7 J.A. Wheeler in *Gravitation* (eds. C.W. Misner, K.S. Thorne and J.A. Wheeler; Freeman 1973) p.1212.

8 A. Einstein to his assistant Ernst Straus, as quoted in *Einstein: A Centenary Volume* (ed. A.P. French; Heinemann 1979) p.128.

9 R.P. Feynman. *The Character of Physical Law* (B.B.C. Publication 1965) pp.124–5.

10 F.J. Dyson, 'Energy in the universe', *Scientific American* (September 1971).

Select Bibliography

In addition to the references given in the text, the following bibliography may be helpful:

Chapter 1 *Science and religion in a changing world*
Many scientists have written about their religious beliefs and many theologians have examined the impact of science on religion. Books falling in the former category are *Science and Faith* by W. Russell Hindmarsh (Epworth 1968); *The Clockwork Image* by Donald MacKay (Inter-Varsity Press 1974); *Chance and Necessity* by Jacques Monod (Random House 1971); *Science and Christian Belief* by C.A. Coulson (Oxford University Press 1955); *Science and Human Values* by Jacob Bronowski (Harper and Row 1965; revised edition) and *Physics, Logic, History* (eds. Allen D. Breck and Wolfgang Yourgrau; Plenum 1970). In the latter category are Stanley Jaki's *Cosmos and Creator* (Scottish Academic Press 1981); A. R. Peacocke's *Science and the Christian Experiment* (Oxford University Press 1971); Thomas Torrance's *Theological Science* (Oxford University Press 1978) and *Divine and Contingent Order* (Oxford University Press 1981) and *Religion and the Rise of Modern Science* by R. Hooykaas (Erdmans 1972). An interesting collection of essays is given in *The Sciences and Theology in the Twentieth Century* (ed. A.R. Peacocke; Oriel 1981).

Chapter 2 *Genesis*
The creation of the universe has been the theme of a large number of popular science books recently. Best known is Steven Weinberg's *The First Three Minutes* (Andre Deutsch 1977; Fontana 1978). See also *The Creation* by P.W. Atkins (Freeman 1981); *Genesis* by John Gribbin

(Dent/Delacorte 1981) and *The Big Bang* by Joseph Silk (Freeman 1980).

The creation, as well as the end of the universe are treated in detail in my book *The Runaway Universe* (Dent 1978; Harper & Row 1978), which emphasizes the significance of the second law of thermodynamics. This is also the theme of Robert Jastrow's *Until the Sun Dies* (Norton 1977). For a straightforward survey of modern cosmology see, for example, *The State of the Universe* (ed. G.T. Bath; Oxford: Clarendon Press 1980); *Cosmology* by E.R. Harrison (Cambridge University Press 1981) or *Modern Cosmology* by D.W. Sciama (Cambridge University Press; second edition 1982).

Many popular expositions exist concerning Einstein's theory of relativity. *Einstein's Universe* by Nigel Calder (B.B.C. publication 1979) gives a superficial but graphic overview. More scholarly, yet still popular works include *The Riddle of Gravitation* by Peter Bergmann (Charles Scribner 1968); *Space, Time and Gravity* by Robert Wald (University of Chicago Press 1977) and the author's *Space and Time in the Modern Universe* (Cambridge University Press 1977). For those interested in actually learning some relativity Sam Lilley's *Discovering Relativity for Yourself* (Cambridge University Press 1981) and Robert Geroch's *General Relativity from A to B* (University of Chicago Press 1978) are for absolute beginners, while Michael Berry's *Principles of Cosmology and Gravitation* (Cambridge University Press 1976) is suitable for those with a little mathematical background.

The subject of infinity is a favourite for mathematical popularization; see for example *Playing with Infinity* by Rozsa Peter (Bell 1961); *Uses of Infinity* by Leo Zippin (Random House 1962), or more recently, *Infinity and the Mind: The Science and Philosophy of the Infinite* by Ruddy Rucker (Harvester 1982). See also 'Mathematics sets about the infinite' by Keith Devlin in *New Scientist, 95*, 162.

Chapter 3 *Did God create the universe?*

An introduction to modern theories of particles and forces, including a discussion of antimatter and particle creation, is given in the author's *The Forces of Nature* (Cambridge University Press 1979) and in Gerald Feinberg's *What is the world made of?* (Doubleday 1977).

A description of unified force theories and proton decay is given by Steven Weinberg in 'The decay of the proton', *Scientific American* (June 1981).

The cosmological argument for the existence of God is reviewed in depth by William Rowe in *The Cosmological Argument* (Princeton

University Press 1975), and by William Lane Craig in *The Cosmological Argument from Plato to Leibniz* (Macmillan 1980). See also *The Cosmological Arguments: a Spectrum of Opinion* by D.R. Burrill (Doubleday-Anchor 1967).

The debate between Bertrand Russell and F.C. Copleston is published in *A Modern Introduction to Philosophy* (eds. P. Edwards and A. Pap; Free Press 1965).

A detailed historical analysis of the problem of a first moment in a cosmological context is given by J.D. North in *The Measure of the Universe* (Oxford: Clarendon Press 1965).

Chapter 4 *Why is there a universe?*

A discussion of the concept of a timeless God is given by Richard Swinburne in *The Coherence of Theism* (Oxford: Clarendon Press 1977). See also *God and Timelessness* by Nelson Pike (Routledge & Kegan Paul 1970) and chapter 8 of Brian Davies's *An Introduction to the Philosophy of Religion* (Oxford University Press 1982).

What is a necessary being? See, for example, Swinburne, op. cit. and a sceptical discussion in John Hosper's *An Introduction to Philosophical Analysis* (Routledge & Kegan Paul, revised edition 1981) chapter 7.

The question of how order can arise from primeval disorder has also been discussed by David Layzer, in 'The arrow of time', *Scientific American* (December 1975). A specific mathematical analysis is contained in 'Entropy in an expanding universe' by Steven Frautschi (to be published).

The role of gravitation in the above problem, and the as yet unsolved problem of gravitational entropy have been discussed in the author's *The Physics of Time Asymmetry* (Surrey University Press/ University of California Press 1974, 1977), and also by Roger Penrose, 'Singularities and time-asymmetry', *General Relativity: an Einstein Centenary Survey* (eds. S.W. Hawking and W. Israel; Cambridge University Press 1979). See also P.C.W. Davies's 'Is thermodynamic gravity a route to quantum gravity?', *Quantum Gravity 2: A Second Oxford Symposium* (eds. C.J. Isham, R. Penrose and D.W. Sciama; Oxford: Clarendon Press 1981).

Chapter 5 *What is life? Holism vs. reductionism*

Provocative views about the nature and origin of life abound. The following selection is just a random sample: *The Life Puzzle* by A.G. Cairns-Smith (Oliver & Boyd 1977); *Life Itself: Its Origin and Nature* by Francis Crick (Macdonald/Simon & Schuster 1982); *The Selfish Gene* by Richard Dawkins (Oxford University Press 1977); *Life Beyond Earth* by Gerald Feinberg and Robert Shapiro (William Morrow 1980); *Lifecloud* by Fred Hoyle and N.C. Wickramasinghe (Dent 1978); Jacques Monod, *Chance and Necessity* (Random House 1971), and *The Origin of Life: Molecules and Natural Selection* by L.E. Orgel (Wiley 1973).

The mechanism-vitalism issue has been dealt with in *The History and Theory of Vitalism* by Hans Driesch (Macmillan 1914) and *Mechanism and Vitalism* by Rainer and Schubert-Soldern (Notre Dame University Press 1962).

The 'Ant fugue' and other lucid expositions of the holism-reductionism issue appear in Douglas Hofstadter's *Gödel, Escher, Bach* (Basic Books 1979). Reductionism in physics comes under strong attack from Fritjov Capra in *The Tao of Physics* (Wildwood House 1975) and *The Turning Point* (Wildwood House 1982). See also *Zen Flesh, Zen Bones* by Paul Reps (Penguin 1971). The oriental-mystical overtones of some of modern holistic physics is also taken up in *The Dancing Wu Li Masters: An Overview of the New Physics* by Gary Zukav (Rider 1979).

Self-organizing chemical and biological systems are studied in depth in *Self-Organization in Non-equilibrium Systems* by G. Nicolis and I. Prigogine (Wiley 1977). The theme is taken up in a broader context in Prigogine's *From Being to Becoming* (Freeman 1980), and in Hermann Haken's *Synergetics* (Springer 1977).

Alien life, intelligent or otherwise, is a favourite topic for both science fiction and popular science, though the difference is not always that clear. Books purporting to be in the latter category include: *The Galactic Club: Intelligent Life in Outer Space* by Ronald Bracewell (Freeman 1974); Feinberg and Shapiro, op. cit.; Hoyle and Wickramasinghe, op. cit; *Interstellar Communication: Scientific Perspectives* (eds. C. Ponnamperuma and A.G.W. Cameron; Houghton-Mifflin 1974); *The Cosmic Connection* by Carl Sagan (Doubleday 1973); *Intelligent Life in the Universe* by I.S. Shkovskii and C. Sagan (Holden-Day 1966) and *We Are Not Alone* by Walter Sullivan (McGraw-Hill 1966). Some consideration about the religious implications of alien life appears in C.S. Lewis's 'Religion and Rocketry', *The World's Last Night and Other Essays* (Harcourt Brace Jovanovich Inc. 1952). See also his *Perelandra* (Macmillan 1944).

Chapter 6 *Mind and soul*

The mind is the central subject of both philosophy and psychology. A good, all round introduction to the many aspects of the mind is Richard Gregory's *Mind in Science* (Weidenfeld & Nicolson 1981), a more informal and speculative survey being *The Mind's I* (eds. D.R. Hofstadter and D.C. Dennett; Harvester/Basic Books 1981), both of which contain extensive bibliographies; see also Dennett's *Brainstorms* (Bradford Books 1978) and the *Oxford Companion to the Mind* (ed. R.L. Gregory; Oxford University Press 1983).

Discussions of artificial intelligence may be found in *Introduction to Artificial Intelligence* by Philip Jackson (Petrocelli Charter 1975), which has a large bibliography; *Machines Who Think* by Pamela McCorduck (Freeman 1979); *Philosophical Perspectives in Artificial Intelligence* (ed. M. Ringle; Humanities Press 1979), and *Artificial Intelligence* by Patrick Winston (Addison-Wesley 1977). See also Douglas Hofstadter's *Gödel, Escher, Bach* (Basic Books 1979).

For a popular introduction to the subject of consciousness, try *Windows on the Mind: Reflections on the Physical Basis of Consciousness* by Eric Harth (Harvester 1982).

A modern defence of dualism is given by Karl Popper and John Eccles in *The Self and its Brain* (Springer 1977) and by H.D. Lewis in *Philosophy of Religion* (The English Universities Press 1965, 1975).

Chapter 7 *The self*

Enigmas of the self and personal identity have long occupied the attentions of philosophers. Recent works include *Problems of the Self* by Bernard Williams (Cambridge University Press 1973); *Personal Identity* (ed. John Perry; University of California Press 1975), and *Self-Knowledge and Self-Identity* by Sydney Shoemaker (Cornell University Press 1963).

An attempt to make Gödel's celebrated theorem comprehensible to those not well versed in formal logic is *Gödel's Proof* by Ernest Nagel and James R. Newman (New York University Press 1958).

The importance of level-distinction in discussion of mind and brain are also emphasized in *Hierarchy Theory: The Challenge of Complex Systems* (ed. H.H. Pattee; George Braziller 1973).

Two books about brain transplants and related identity confusions are Perry, op. cit. and *The Identities of Persons* (ed. A.O. Rorty; University of California Press 1976).

Chapter 8 *The quantum factor*

Popularizations of the quantum theory are rare. I have tried my best in *Other Worlds* (Dent 1980). The holistic aspects are emphasized in Bohm's *Wholeness and the Implicate Order*. For a conventional introduction see, for example, *The Strange Story of the Quantum* by Banesh

Hoffmann (Dover, second edition 1959). For those with technical knowledge a good survey of the epistemological problems is Bernard d'Espagnat's *Conceptual Foundations of Quantum Mechanics* (Benjamin 1971). Easier to start with is his 'Quantum theory and reality', *Scientific American* (November 1979).

Among the classics by the founders of the theory themselves are *Atomic Theory and the Description of Nature* by Niels Bohr (Cambridge University Press 1934); *The Physical Principles of Quantum Theory* (University of Chicago Press 1930); *Physics and Philosophy* (Harper & Row 1958) and *The Physicist's Conception of Nature* (Hutchinson 1958) by Werner Heisenberg, and *Natural Philosophy of Cause and Chance* by Max Born (Oxford University Press 1949).

A useful collection of ideas appears in *Quantum Theory and Beyond* (ed. T. Bastin; Cambridge University Press 1971).

The experiment of Aspect et. al. is described in *Physical Review Letters* 49, 1804 (1982).

John Wheeler's 'delayed choice' experiment is described in his article 'Beyond the black hole', *Some Strangeness in the Proportion: A Centennial Symposium to celebrate the Achievements of Albert Einstein* (ed. H. Woolf; Addison-Wesley 1980).

The analysis of the quantum measurement problem by J. von Neumann is given in his book *Mathematical Foundations of Quantum Mechanics* (Princeton University Press 1955).

The Everett interpretation of the quantum theory is developed in full detail in *The Many-Worlds Interpretation of Quantum Mechanics* by B.S. DeWitt and N. Graham (Princeton University Press 1973). For an easier introduction, see DeWitt's article 'Quantum mechanics and reality', *Physics Today* (September 1970).

Bell's inequality was first proved in *Physics 1*, 195 (1964).

Chapter 9 *Time*

As one of the perennially fascinating topics in physics, philosophy and theology the literature on time is immense. For the basic physics of time at an elementary level of exposition try my *Space and Time in the Modern Universe* (Cambridge University Press 1977); G.J. Whitrow's *The Natural Philosophy of Time* (Nelson 1961), or any introductory book on the theory of relativity (see bibliography for Chapter 2).

Books with a special emphasis on the physics and philosophy of the 'arrow' of time are *The Physics of Time Asymmetry* by P.C.W. Davies (Surrey University Press/University of California Press 1974, 1977), which contains many references; *The Nature of Time* (ed. T. Gold;

Cornell University Press 1967); *The Direction of Time* (University of California Press 1956, 1971) and *The Philosophy of Space and Time* (Dover 1958) by Hans Reichenbach.

For an introduction to the general philosophical and theological aspects of time see, for example, *The Concepts of Space and Time* (ed. M. Capek; Reidel 1976); *Philosophy of Space and Time* by M. Capek (Dover 1958); *The Genesis and Evolution of Time* by J.T. Fraser (University of Massachusetts Press 1982); *The Study of Time* (eds. J.T. Fraser et. al.; Springer) Vol. I (1972), Vol. II (1975), Vol. III (1978); *Philosophy of Time* (eds. E. Freeman and W. Sellars; Open Court 1971); *Time, Reduction and Reality* (ed. R. Healey; Cambridge University Press 1981); *God and Timelessness* by Nelson Pike (Routledge & Kegan Paul 1970); *Space, Time and Spacetime* by L. Sklar (University of California Press 1974); *Space and Time* (ed. J.J.C. Smart; Macmillan 1964), and *Space and Time* by Richard Swinburne (Macmillan 1968).

Problems and paradoxes concerning the 'now' are developed by A. Grünbaum in *Philosophical Problems of Space and Time* (Alfred A. Knopf 1964); J. McT. E. MacTaggart's attempt to reject the time concept entirely is described in 'The unreality of time', *Mind 18*, 457 (1908).

For a more entertaining study of some of these topics John Gribbins's *Timewarps* (Dent 1979) should appeal.

For further reading on the subject of black holes, see under Chapter 13.

Chapter 10 *Freewill and determinism*

A lucid discussion of freedom, fatalism, determinism and indeterminism is given in Hospers op. cit.

A number of books have been devoted specifically to the subject; for example, *Free-will and Determinism* (ed. B. Berofsky; Harper & Row 1966); *Determinism and Freedom in the Age of Modern Science* (ed. S. Hook; New York University Press 1957), and *Freedom and Determinism* (ed. K. Lehrer; Random House 1965).

The importance of God's freedom of action is stressed in Swinburne's *The Coherence of Theism* (Oxford: Clarendon Press 1977).

Chapter 11 *The fundamental structure of matter*

Introductory surveys of modern particle physics include *The Key to the Universe* by Nigel Calder (Viking Press 1977); *From Atoms to Quarks: An Introduction to the Strange World of Particle Physics* by J.S. Trefil (Charles Scribner 1980) and *The Particle Play* by J.C. Polkinghorne (Freeman 1980), as well as the books by Feinberg and the author given in the bibliography of Chapter 3.

Particle physics and the theory of fundamental forces is so fast moving that the best introduction is from the regular review articles carried by *Scientific American*. I recommend the following: 'Quarks with colour and flavour' by S.L. Glashow (October 1975); 'The confinement of quarks' by Y. Nambu (November 1976); 'Unified theories of elementary particle interactions' by S. Weinberg (July 1974). Taking the matter further, to include the so-called grand unified theories, is 'Unified theory of elementary particles and forces' by Howard Georgi (April 1981).

The ultimate theory — supergravity — is reviewed by D.Z. Freedman and P. von Nieuwenhuizen in 'Supergravity and the unification of the laws of physics' (February 1978). Some of these articles (and others) are reprinted in *Particles and Fields* (ed. W.J. Kaufmann III; Freeman 1980).

Finally, some recent thinking about the application of the above ideas to the very early universe is summarized by Edward Kolb and Michael Turner in 'The early universe', *Nature 294*, 521 (1981).

The holistic aspect of particles being 'made up' of other particles is emphasized in Capra's *The Tao of Physics*.

Chapter 12 *Accident or design?*

A catalogue of apparent 'accidents' or 'coincidences' in nature, and possible interpretations thereof, are reviewed in detail in my book *The Accidental Universe* (Cambridge University Press 1982) and in *The Anthropic Principle* by John Barrow and Frank Tipler (Oxford University Press, to appear) which also contains a fascinating and in-depth historical analysis.

A summary of 'order and disorder in the universe' is given under that title by the author in *The Great Ideas Today* (Encyclopedia Britannica 1979). See also the author's *The Runaway Universe* (Dent 1978; Harper & Row 1978).

Discussion of Poincaré cycles can be found in Reichenbach's *The Direction of Time*, (University of California Press 1956, 1971) and in the author's *The Physics of Time Asymmetry* (Surrey University Press/University of California Press 1974, 1977).

The cyclic universe model was analysed by R. Tolman in *Relativity, Thermodynamics and Cosmology* (Oxford: Clarendon Press 1934) and became the basis of John Wheeler's 'reprocessing' universe discussed by him in, for example, *Gravitation* by C.W. Misner, K.S. Thorne and J.A. Wheeler (Freeman 1973) chapter 44.

A number of discussions of 'The Anthropic Principle' have appeared in the recent literature: 'The anthropic principle and the structure of the physical world' by B.J. Carr and M.J. Rees, *Nature 278*, 605 (1979) and 'Physical laws and the numerical values of the fundamental constants' by I.L. Rozental in *Soviet Physics* (Uspekhi) *23*, 296 (1980) are excellent introductions to the physics of the subject. Some of the philosophical aspects are covered by John Leslie in 'Anthropic principle, world ensemble and design' in *American Philosophical Quarterly 19*, 141 (1982). See also 'Did God create the universe?' by John Bowker in *The Sciences and Theology in the Twentieth Century* (ed. A.R. Peacocke; Oriel 1981) and, of course, Barrow and Tipler op. cit. An introductory summary can be found in 'The anthropic principle' by George Gale in *Scientific American* (December 1981).

Chapter 13 *Black holes and cosmic chaos*

Black holes are a favourite topic for popularization, but care should be exercised: the subject has often been badly handled. Careful yet fascinating discussions have been given by Larry Shipman in *Black Holes, Quasars and the Universe* (Houghton-Mifflin 1976); Iain Nicolson in *Gravity, Black Holes and the Universe* (David & Charles 1981) and William Kaufmann in *Black Holes and Warped Spacetime* (Freeman 1979), as well as most popular books on relativity (see bibliography for Chapter 2). I have also discussed the subject in depth in *The Edge of Infinity* (Dent 1981).

For a discussion of the concept of gravitational entropy and the importance of the gravitational field for the subject of order and disorder in the universe, see the author's *The Physics of Time Asymmetry* (Surrey University Press/University of California Press 1974, 1977). In recent years the subject has been developed and clarified greatly by Roger Penrose. See, for example, his 'Singularities and time-asymmetry' in *General Relativity: an Einstein Centenary Survey* (eds. S.W. Hawking and W. Israel; Cambridge University Press 1979). This volume also contains an article by R.H. Dicke and P.J.E. Peebles entitled 'The big bang cosmology — enigmas and nostrums' which discusses the horizon problem mentioned on page 180, and related topics.

The difficulties concerning overproduction of heat by the dissipation of primeval turbulence were emphasized by J.D. Barrow and R.A. Matzner, 'The homogeneity and isotropy of the universe' in *Monthly Notices of the Royal Astronomical Society 181*, 719 (1977). See also Barrow's 'Quiescent cosmology', *Nature 272*, 211 (1978). John Barrow has pointed out to me that even if protons do decay, some primeval irregularities could still leave an imprint on the later universe, viz. spatial inhomogeneities. This is discussed in his paper with M.S. Turner in *Nature 291*, 469 (1981).

Freeman Dyson's discussion of the catastrophic consequences attendant on a slight change in the strength of the nuclear force was given in his article entitled 'Energy in the universe' in *Scientific American* (September 1971).

Teleological arguments for the existence of God have a long history of assent and dissent. Aquinas, adapting Aristotelean ideas of activity, developed the notion of final cause as evidence for a divine purpose in the world. However, it is much later, for example in the work of Paley and F.R. Tennant's *Philosophical Theology* (Cambridge University Press 1969; first published 1928) that the argument becomes fully developed. It was attacked by I. Kant in his *Critique of Pure Reason* (trans. J.M.D. Meikeljohn; Dent 1969; first published 1781) and D. Hume in his *Dialogues Concerning Natural Religion* (ed. H.D. Aiken; Hafner 1969; first published 1779).

For a modern discussion, see *The Argument from Design* by Thomas McPherson (Macmillan 1972), and Richard Swinburne's article of the same title in *Philosophy 43*, 200 (1968). Recent attempts at reformulating teleology as 'teleonomy' are described by Bowker, op. cit.

Chapter 14 *Miracles*

In chapter X of his *Enquiry Concerning Human Understanding* (ed. L.A.S. Bigge; Greenwood Press 1980; first published 1758) Hume provides perhaps the most famous critical assessment of the subject of miracles.

For a modern appraisal, see *Interpreting the Miracles* by R.H. Fuller (S.C.M. 1966); *The Concept of Miracle* by R. Swinburne (Macmillan 1970) and chapter 11 of Brian Davies's *An Introduction to the Philosophy of Religion* (Oxford University Press 1982).

A sceptical survey of the so-called paranormal by a formerly convinced scientist is to be found in *Science and the Supernatural* by J.G. Taylor (M.T. Smith 1980).

Chapter 15 *The End of the Universe*

I have dealt with this theme in greater detail in *The Runaway Universe* (Dent 1978; Harper & Row 1978). There are a number of recent articles and papers devoted to the subject: for example, 'The ultimate fate of the universe' by J.N. Islam, *Sky and Telescope 57*, 13, (January 1979);

'Eternity matters' by D.N. Page and M.R. McKee, *Nature 291*, 44 (1981) and 'Eternity is unstable' by J.D. Barrow and F.J. Tipler, *Nature 276*, 453 (1978). Barrow has argued that although cosmic matter may degenerate to a heat death, distortions in the cosmic gravitational fields could be continually sustained by slight irregularities in the expansion rate of the universe. These distortions could be used as an unending source of free energy. See also 'The collapse of the universe: an eschatological study' by M.J. Rees, *Observatory 89*, 183 (1969).

The intriguing issue of whether intelligent life can survive in an ever-expanding universe approaching the heat death is dealt with in detail by Freeman Dyson in 'Time without end: physics and biology in an open universe', *Reviews of Modern Physics 51*, 447 (1979).

An introduction to Maxwell's demon is given by W. Ehrenberg, 'Maxwell's demon', *Scientific American* (November 1967). L. Szilard's paper 'On the reduction of entropy of a thermodynamic system caused by intelligent beings' was published in *Zeitschrift für Physik 53*, 840 (1929). The connection between information and entropy in 'demonology' was not, however, fully developed until Leon Brillouin's paper 'Maxwell's demon cannot operate', *Journal of Applied Physics 22*, 334 (1951).

Chapter 16 *Is the universe a 'free lunch?'*

A number of physicists have explored the idea that the universe might be some sort of quantum fluctuation. In the less bold versions, the fluctuation is from empty, flat spacetime. See, for example, 'Is the universe a vacuum fluctuation?' by E.P. Tryon, *Nature 246*, 396 (1973); 'The creation of the universe as a quantum phenomenon' by R. Brout, F. Englert and E. Gunzig, *Annals of Physics 115*, 78 (1978); 'Origin of the universe as a quantum tunneling event' by D. Atkatz and H. Pagels, *Physical Review D 25*, 2065 (1982).

In the bolder versions, the fluctuation is from nothing at all. This is the theme of a paper by the Soviet astrophysicist Ya. B. Zeldovich in *Soviet Astronomy Letters* 7, 579 (1981). See also L.P. Grishchuk and Ya. B. Zeldovich 'Complete cosmological theories', *The Quantum Theory of Space and Time* (ed. M.J. Duff and C.J. Isham; Cambridge University Press 1982) and A. Vilenkin 'Creation of the universe from nothing', *Physics Letters B117*, 25 (1982).

A review of the inflationary universe scenario may be found in 'The inflationary universe — birth, death and transfiguration' by J.D. Barrow and M.S. Turner, *Nature 298*, 801 (1982).

Chapter 17 *The physicist's conception of nature*

The best way to learn how physicists think about the world is by talking to them. A guide is therefore provided by *A Question of Physics: Conversations in Physics and Biology* (eds. P. Buckley and F. Peat; Routledge & Kegan Paul 1979).

In September 1972 a conference was held in Trieste in honour of Paul Dirac's seventieth birthday. Some of the papers can be found in the book *The Physicist's Conception of Nature* (ed. J. Mehra; Reidel 1973). See also Heisenberg's (1958) book of the same title.

Another celebratory event was Einstein's centenary in 1979, from which a number of volumes on the 'conception' theme emerged. For example H. Woolf, (ed.) *Some Strangeness in the Proportion: A Centennial Symposium to celebrate the Achievements of Albert Einstein* (Addison-Wesley 1980), has several broad-ranging conceptual articles as well as valuable historical material about Einstein's conception of nature (not always shared by his colleagues).

Finally, an excellent layperson's introduction to the physicist's approach is *The Character of Physical Law* by Richard Feynman (B.B.C. Publications 1965), based on his radio talks.

Not everybody thinks the physicists have got it right. For a critical appraisal of some earlier thinking see L.S. Stebbing's *Philosophy and the Physicists* (Pelican 1944).

A description of John Conway's game 'Life' appears in *Winning Ways* by E.R. Berlekamp, J.H. Conway and R.K. Guy (Academic Press 1982, Vol. II).

Index

ABOUT THE AUTHOR

Dr. Paul Davies is Professor of Theoretical Physics, University of Newcastle-upon-Tyne, England. From 1970 to 1972, he was visiting fellow at the Institute of Astronomy, Cambridge, and from 1972 to 1980, he was lecturer in applied mathematics at King's College, University of London. He writes internationally for science magazines and journals, including *Nature, New Scientist, The Economist,* and *The Sciences,* and he frequently contributes to science broadcasts. He is the author of *The Edge of Infinity, Other Worlds, The Physics of Time Asymmetry, Space and Time in the Modern Universe, The Runaway Universe, The Forces of Nature,* and *The Search for Gravity Waves.*